Lecture Notes in Mathematics

A collection of informal reports and seminars
Edited by A. Dold, Heidelberg and B. Eckmann, Zürich

145

Eduardo J. Dubuc
University of Chicago, Chicago, IL/USA

Kan Extensions in Enriched Category Theory

Springer-Verlag
Berlin · Heidelberg · New York 1970

P R E F A C E

Category Theory is rapidly coming of age as a Mathematical discipline. In this process it now appears that a central role will be played by the notion of an enriched category. These categories, with hom-sets in a closed category,--in particular, in a symmetric monoidal closed category--seemed initially very complex and difficult to manage effectively. However, independent work by various experts--Yoneda, Linton, Bénabou, Eilenberg-Kelly, Lambek, Bunge, Ulmer, Gray, Palmquist, and others--has considerably improved the situation. A vital step was the discovery of the proper use of tensors, cotensors, and Kan extensions for enriched categories (A discovery made simultaneously and independently by Bénabou and by Kelly with Day). As a result, an efficient presentation of enriched categories is now possible.

This paper by Dubuc collects all these ideas in a compact exposition which makes this efficiency very clear--and which also serves as a basis for Dubuc's own original contributions. I have, therefore, recommended to the editors of the <u>Lecture Notes</u> series the rapid publication of this paper, to provide easy access to this foundation for future development.

Saunders Mac Lane

TABLE OF CONTENTS

INTRODUCTION

The original purpose of this paper was to provide suitable enriched completions of small enriched categories. We choose as a precise meaning for the words "enriched category" the notion of V-category, where $\mathbb{V}$ is any given (fixed) symmetrical monoidal closed category ([2]), abbreviated closed category (see [3], introduction). In introducing the notation we recall the above notions, but the reader so inclined can perfectly well go through this paper thinking "category", "functor" every time he sees the words "V-category", "V-functor", (and etc.).

We found it necessary to set up an appropriate background in which facts about an enriched world are stated, and in doing so we (comfortably) put ourselves in a non-autonomous treatment whose basic guidelines can be subsumed in the questions: "From which minimal set of basic facts about the set-based world can we deduce the existence of completions of small categories?", "Which of those facts, literally or suitable translated, are still true in the V-world, and what conditions should be imposed on $\mathbb{V}$ in order to rescue all of them?" It is clear that given any result of set-based category theory these two questions can be used as guidelines for an investigation of the enriched worlds. We hope that in this paper it is shown that the result concerning the existence of completions for a small category, rich enough in different notions but yet simple, leads to sufficient knowledge of the V-world of nature, knowledge which is indispensible to achieve the desirable formulation and development of an appropriate autonomous (axiomatic) foundation of the Category of V-categories (and of V itself), in a way similar to the one developed in the pioneer work of Lawvere for the ordinary set-based world. We also hope that the techniques employed in this paper, techniques that rest on an intensive use of Kan extensions and which are perfectly suited for such an autonomous treatment, have as well (because of their simplicity and economy) their own interest even when applied to obtain the known ordinary results, bringing a better understanding of the reasons why those results hold. On the other hand, due to the proliferation of closed categories interesting for mathematical practice (see for example Bunge [7] where a list of some of them is provided) this paper should provide a common setting for many different constructions and results in mathematics.

In order to have a handy reference in pre-section 0 we took from Kelly [3] the basic results concerning V-adjunctions. Talking about only one side of the duality, the unique completeness concept (right Kan extension) of ordinary set-based category splits into four different (and independent) ones in the V-context. This is a clear imperfection of this non-autonomous treatment, but a careful reading of the paper seems to indicate that only two of them are essentially needed, namely, the one of cotensor and the one of right Kan extension . In I.1, I.2 and I.3 we introduce V-limits, cotensors and ends, and most of the definitions and results were taken from the summary papers of Day-Kelly [1], [3], except for our careful treatment of the concepts applied to V-functors, where a distinction appears between V-functors which satisfy the universal property and V-functors which pointwise satisfy the universal property. The later ones are characterized by the fact that the universal property is preserved by the representable functors of the codomain category. This treatment also serves to fill a gap in ordinary category theory, that, if probably known by many authors, has to my best knowledge never been written down on paper. In I.4 we make a parallel treatment of right Kan extensions, where we set the properties which enable us to use Kan extensions as the single major tool through all the paper. We also give a formal criterion for the existence of V-left adjoint (Benabou [6]) and the V-version of the well known classical Kan formula, (which we took from [3], where a proof of a stronger result is given which applies only in the small and complete case).

We thought that the best way of introducing the relevant concepts of generator, generating functor and dense functor, as well as the best way to do the necessary constructions needed in the completion of a small category (Lambeck [4]) was by means of the use of monads (triples). This technique has also the advantage that it can be generalized to the V-context without any further complications. Central is the concept of codensity monad, which we took from Linton [9], and which we introduce here as a particular right Kan extension.

We found it also necessary to use a fair number of properties of monads in the V-context (V-monads), and so in Chapter II we put Kan extensions to work in order to develop that part of the theory of V-monads that we use later in Chapter III and IV.

It should be noticed that Chapter II (as well as the rest of the paper) is written without recourse to ordinary set-based results, and thus, when applied to this case, it provides new techniques (different proofs) for achieving these results. These techniques should be called "Kan extensions techniques". For example; in II.1, we observe that the general Semantics-Structure adjointness is just the Kan extension universal property of the codensity V-monad.

In III.1 we develop general properties of V-continuous V-functors; and in doing so we have follow as a basic guideline the paper of Lambeck [4], some of whose propositions are here literally translated into the V-context. In III.2 we assume for the first time (except for the case of cotensors) the existence of completeness concepts in the categories that we work with, and we develop the special properties of V-complete V-categories. Central in this section is the V-Special Adjoint Functor Theorem, which we deduce as a corollary from properties of Kan extensions. The proof of this theorem suggests that the relevant property of cogenerators is not that of producing solution sets, but the fact that just by definition the unit of the associated codensity monad is a monomorphism. For example, in the V-context, a V-cogenerator will in general not be a real cogenerator, (the category $\mathbb{1}$ is a $\mathbb{C}$at-generator for $\mathbb{C}$at), but the special Adjoint Functor Theorem still holds.

In III.3 the concept of V-completion is introduced, and what we mean exactly by completion is explained. This has also been taken from Lambeck [4]. The reader interested in the problem of completions should (of course) consult the (fundamental) work of Isbell, that, because of its different language we have found difficult to introduce (or refer to) here. We give two different methods of constructing a V-completion for a small V-category. The first is just an adequate V-version of the completion obtained in Lambeck [4], and in order to achieve it we use a technique based on a straightforward use of V-monads. The second use the construction of a tower of categories and functors by means of limits of (small) chains of categories obtained as a result of an iterated construction of categories of algebras over a monad. The author first learned of this construction in a lecture given by Tierney in the Midwest Category Seminar at Urbana Illinois, where Tierney also suggested its possible usefulness in the construction of completions. After that lecture the author and Tierney himself discovered independently the crucial fact that the whole tower (a large chain of categories) has a limit which is a locally small category, and that the corresponding monad in that limit is the identity. Here we present a V-version of all these facts and we use them to provide a V-completion.

In IV.1 the construction of the V-category of V-functors and V-natural transformations is made in exactly the same way as in Day-Kelly [1], and in IV.2 we use the V-Yoneda embeddings as the starging data upon which we apply the process of construction of V-completions developed in III.3. Finally in IV.3 we take the definitions of corealization and cosingular functors from Applegate, Tierney [5], we develop the additional features of the tower constructed in III.3 under the presence of V-functor categories and we give a comparison of the two completions.

An appendix is given where we find conditions on a closed category $\mathbb{V}$ which imply that in the V-world cotensors in V-categories with limits are real limits.

The author hopes that it will not be totally incorrect to say that this paper is a testimony of two basic <u>mathematico-philosophical principles</u>. <u>First</u>, "the relevant properties of mathematical objects are those which can be stated in terms of their abstract structure rather than in terms of the elements which the objects were thought to be made of (Lawvere)" coupled with "the relevant facts of category theory hold because of formal interconnections between the concepts involved rather than because of their substantial content (which is none)".

This, because of that peculiar characteristic of the mind which leads every human being to the convinction that abstract ideas are real, can be pushed forward (extrapolated) into a simple purely philosophical principle, namely, "substance <u>is</u> form". <u>Second</u>, "everything in mathematics that can be categorized is trivial (Freyd)" which should be understood. "Category Theory is good ideas rather than complicated techniques."

* * *

We omit in the text all remarks concerning the uniqueness up to isomorphisms of concepts defined by means of universal properties, thus this fact should be present in the mind of the reader, especially because of our repeated use of the article "the" instead of the article "a".

We often denote isomorphisms by the same letter in both directions (especially in the case of adjunctions).

There are three kinds of statements in this paper, the ones which hold without any assumptions in $\mathbb{V}$ are headed in the usual way (<u>Proposition ...</u>). The ones which hold only when $\mathbb{V}$ has equalizers are headed with one black "•" preceeding them (<u>• Proposition ...</u>). The ones which hold only when $\mathbb{V}$ is a complete category (i.e., when it has (small) limits) are headed with two black "•" preceeding them (<u>•• Proposition...</u>).

Besides the few simplifying conventions mentioned above, we believe that other peculiarities that may have escaped the attention of the author will not lead to any confusion for the reader.

The rest of the notation is introduced in the text.

* * *

My thanks to M. Tierney and Ross Street for many conversations about the material presented in this paper, my special thanks to Professor A. P. Calderon who was responsible for my coming to the University of Chicago, and, my heartful thanks to Professor Saunders MacLane for help, mathematical and non-mathematical, without which this paper could not have been written. For his continued teaching, criticism and encouragement the value of my close contact with him during the past three years is impossible to overestimate.

E. J. D. December 1969

Terminology

A category $\mathbb{V}$ is monoidal if it has an associative tensor product $\mathbb{V} \times \mathbb{V} \xrightarrow{\otimes} \mathbb{V}$ with a unit $I \in \mathbb{V}$ (that is, $I\otimes - \approx id$ and $-\otimes I \approx id$) and coherence.

A monoidal category $\mathbb{V}$ is symmetric if for every $V, W \in \mathbb{V}$, $V \otimes W \approx W \otimes V$, (natural) and coherence.

We call the functor $\mathbb{V} \xrightarrow{\mathbb{V}_o(I,-)} \mathbb{S}$ the base functor.

If $\mathbb{V}$ is a monoidal category, a V-category $\mathbb{A}$ is a class of objects A, B, C, ..., for any two A, B, a V-object "between" them, $\mathbb{A}(A,B) \in \mathbb{V}$, for any three A, B, C a "composition", that is, a map $\mathbb{A}(A,B) \otimes \mathbb{A}(B,C) \xrightarrow{o} \mathbb{A}(A,C)$ and for any A an "identity", that is, a map $I \xrightarrow{i} \mathbb{A}(A,A)$ in $\mathbb{V}$. This data is subject to the requirement that "o" be associative and "i" be a unit for "o".

If $\mathbb{A}$ is a V-category, a category (with the same class of objects) is obtained by defining $\mathbb{A}_o(A,B) \in \mathbb{S}$, $\mathbb{A}_o(A,B) = \mathbb{V}_o(I, \mathbb{A}(A,B))$. This category is usually called the "underlying" category of $\mathbb{A}$. By an abuse of language we consider $\mathbb{A}$ to be at the same time a V-category and a category. Thus, we use only one notational symbol, $\mathbb{A}$. $\mathbb{A}_o(AB)$ denotes the set of morphisms in $\mathbb{A}$ between A and B.

$\mathbb{A}(A,B)$ becomes a functor $\mathbb{A}^{op} \times \mathbb{A} \xrightarrow{\mathbb{A}(-,-)} \mathbb{V}$, $(A,B) \rightsquigarrow \mathbb{A}(A,B)$, and so for every (fixed) $A \in \mathbb{A}$ there is a functor $\mathbb{A} \xrightarrow{\mathbb{A}(A,-)} \mathbb{V}$ whose action on a morphism $B \xrightarrow{f} B'$ we often denote by $\mathbb{A}(A,B) \xrightarrow{\mathbb{A}(\square,f)} \mathbb{A}(A,B')$ and dually

A V-functor between two V-categories; $\mathbb{A} \xrightarrow{F} \mathbb{B}$ is a function from the class of objects of $\mathbb{A}$ to the class of objects of $\mathbb{B}$, and for any two $A, B \in \mathbb{A}$ a map $\mathbb{A}(A,B) \xrightarrow{FAB} \mathbb{B}(FA, FB)$ in $\mathbb{V}$, preserving the units and the composition.

If $\mathbb{A} \xrightarrow{F} \mathbb{B}$ is a V-functor, a functor, which we also denote by $\mathbb{A} \xrightarrow{F} \mathbb{B}$ is obtained by defining

$$\mathbb{A}_o(A,B) = \mathbb{V}_o(I, \mathbb{A}(AB)) \xrightarrow{\mathbb{V}_o(I, F_{AB})} \mathbb{V}_o(I, \mathbb{B}(FA,FB)) = \mathbb{B}_o(FA,FB) .$$

This functor is usually called the "underlying" functor of F. By an abuse of language, we consider F to be at the same time a V-functor and a functor.

Since $\mathbb{V}_o(I, F_{AB}) = \mathbb{V}_o(I, G_{AB})$ does not imply $F_{AB} = G_{AB}$, it is clear that different V-functors can be equal as functors, except when I is a generator, that is, when the base functor is faithful.

A V-natural transformations between two V-functors $F \xRightarrow{\varphi} G$ is a family of maps $FA \xrightarrow{\varphi A} GA$ such that for every pair of objects the diagram:

$$\begin{array}{ccc} \mathbb{A}(A,B) & \xrightarrow{F} & \mathbb{B}(FA,FB) \\ \Big\downarrow{\scriptstyle G} & & \Big\downarrow{\scriptstyle \mathbb{B}(\square,\varphi B)} \\ \mathbb{B}(GA,GB) & \xrightarrow{\mathbb{B}(\varphi A,\square)} & \mathbb{B}(FA,GB) \end{array} \qquad \text{commutes}$$

A V-natural transformation is a natural transformation but not vice-versa, except when I is a generator.

A <u>closed category $\mathbb{V}$</u> is a symmetrical monoidal category $\mathbb{V}$ such that for any object $\mathbb{V} \in \mathbb{V}$ the functor $\mathbb{V} \xrightarrow{-\otimes V} \mathbb{V}$ has a right adjoint $\mathbb{V} \xrightarrow{\mathbb{V}(V,-)} \mathbb{V}$.

By the aid of the adjunction isomorphism maps can be defined $\mathbb{V}(V,W) \otimes \mathbb{V}(W,X) \longrightarrow \mathbb{V}(V,X)$ and $I \longrightarrow \mathbb{V}(V,V)$ making of $\mathbb{V}$ a V-category. So there is a functor $\mathbb{V}^{op} \times \mathbb{V} \xrightarrow{\mathbb{V}(-,-)} \mathbb{V}$ and for every V, the functor $\mathbb{V}(-,V)$ is adjoint on the right to itself.

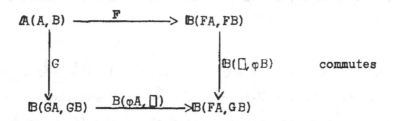

For any V-category $\mathbb{A}$ and object $A \in \mathbb{A}$ the functors $\mathbb{A} \xrightarrow{\mathbb{A}(A,-)} \mathbb{V}$ and $\mathbb{A}^{op} \xrightarrow{\mathbb{A}(-,A)} \mathbb{V}$ become V-functors and there is a V-functor $\mathbb{A}^{op} \otimes \mathbb{A} \xrightarrow{\mathbb{A}(-,-)} \mathbb{V}$ (where $\mathbb{A}^{op} \otimes \mathbb{A}$ is the V-category whose class of objects is the cartesian product (Objects of $\mathbb{A}$) x (Objects of $\mathbb{A}$) with V-structure given by: $\mathbb{A}^{op} \otimes \mathbb{A}((A,B),(A',B')) = \mathbb{A}(A',A) \otimes \mathbb{A}(B,B')$. There is a (canonical) functor $\mathbb{A}^{op} \times \mathbb{A} \longrightarrow \mathbb{A}^{op} \otimes \mathbb{A}$.

When $\mathbb{A} = \mathbb{V}$, for every $V \in \mathbb{V}$ the V-functor $\mathbb{V} \xrightarrow{\mathbb{V}(V,-)} \mathbb{V}$ is V-right adjoint to the V-functor $\mathbb{V} \xrightarrow{-\otimes V} \mathbb{V}$ and the V-functor $\mathbb{V}(-,V)$ is V-adjoint on the right with itself (for the definition of V-adjointness see 0)

A morphism in $\mathbb{V}$ which is both a monomorphism and an epimorphism would not be in general an isomorphism. Hence V-full and V-faithful (clear definition) V-functors need not be such that the maps $\mathbb{A}(A,B) \xrightarrow{F_{AB}} \mathbb{B}(FA,FB)$ are isomorphisms. A V-functor such that F_{AB} is an isomorphism is called V-full-and-faithful.

V-monomorphisms

In ordinary category theory monomorphisms in a category $\mathbb{A}$ are maps $A \xrightarrow{m} B$ in $\mathbb{A}$ such that for every object $A \in \mathbb{A}$ $\mathbb{A}_0(A,m)$ is an injective function in sets. This property, when $\mathbb{A}$ is a V-category, does not imply that $\mathbb{A}(A,m)$ is a monomorphism in $\mathbb{V}$. Since this latter fact is essential for the concept of monomorphisms, we are forced to give:

Definition 0.1

Given a V-category $\mathbb{A}$, a morphism $B \xrightarrow{m} C$ in $\mathbb{A}$ is a V-monomorphism if for every $A \in \mathbb{A}$ the morphism $\mathbb{A}(A, B) \xrightarrow{\mathbb{A}(A,m)} \mathbb{A}(A,C)$ is a monomorphism in $\mathbb{V}$.

Since the base functor sends monomorphism into monomorphism it follows that V-monomorphisms are monomorphisms. The functors $\mathbb{V}(V,-)$ send monomorphisms into monomorphisms into monomorphisms and so in $\mathbb{V}$ itself both concepts coincide. Hence, for any $\mathbb{A}$, the representable functors send V-monomorphisms into V-monomorphisms. The usual formal properties of monomorphisms hold in the V-content, for example: If $A \to B \to C$ is a V-monomorphism then so is $A \to B$. If both $A \to B$ and $B \to C$ are V-monomorphisms then so is $A \to B \to C$. If a morphism splits, then it is a V-monomorphism .

Similarly there is the dual notion of V-epimorphisms.

V-Adjunctions

Definition 0.2

Let $\mathbb{A}$, $\mathbb{B}$ be any two V-categories and $\mathbb{A} \xrightarrow{F} \mathbb{B}$, $\mathbb{B} \xrightarrow{G} \mathbb{A}$ any two V-functors. We say that F is V-left adjoint to G and G V-right adjoint to F if there are V-natural transformations $FG \overset{\epsilon}{\Longrightarrow} id_{\mathbb{B}}$, $GF \overset{\eta}{\Longleftarrow} id_{\mathbb{A}}$ such that the following diagrams commute:

$$(1) \qquad \begin{array}{ccc} G & \overset{\eta G}{\Longrightarrow} & GFG \\ & \searrow\!\!\!\!\searrow & \Downarrow G\epsilon \\ & & G \end{array} \quad , \qquad \begin{array}{ccc} F & \overset{F\eta}{\Longrightarrow} & FGF \\ & \searrow\!\!\!\!\searrow & \Downarrow \epsilon F \\ & & F \end{array} \quad .$$

We denote this situation by $(\epsilon, \eta)\colon F \dashv_V G$, and call F, G a pair of V-adjoint functors; (ϵ, η) a V-adjunction; ϵ the counit and η the unit and we refer to equations (1) as the triangular equations.

There is a bijection between V-adjunctions ϵ, η and V-natural isomorphisms $\mathbb{B}(F(-),-) \underset{\approx}{\overset{\theta}{\Longrightarrow}} \mathbb{A}(-, G(-))$. We also call the latter a V-adjunction and denote it by $\theta\colon F \dashv_V G$, with θ the adjunction isomorphism, denoted by the same letter in both directions. $\theta \circ \theta = id$.
The bijection is given by means of the following definitions:

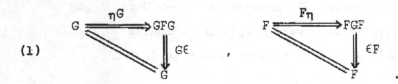

(2) $\theta_{A,B} = (\mathbb{B}(FA, B) \xrightarrow{G_{FA,B}} \mathbb{A}(GFA, GB) \xrightarrow{\mathbb{A}(\eta A, GB)} \mathbb{A}(A, GB))$,

(3) $\theta_{A,B} = (\mathbb{B}(FA, B) \xleftarrow{\mathbb{B}(FA, \epsilon B)} \mathbb{B}(FA, FGB) \xleftarrow{F_{A,GB}} \mathbb{A}(A, GB))$,

(4) $\eta_A = (I \xrightarrow{i_{FA,FA}} \mathbb{B}(FA, FA) \xrightarrow{\theta_{A,FA}} \mathbb{A}(A, GFA))$,

(5) $\epsilon B = (I \xrightarrow{i_{GB,GB}} \mathbb{A}(GB, GB) \xrightarrow{\theta_{GB,B}} \mathbb{B}(FGB, B))$.

At the level of sets we write: $\theta_o \dfrac{FA \longrightarrow B}{A \longrightarrow GB}$,

and, equivalently, we have:

(4), (5) $\eta_A = \theta_o(FA \xrightarrow{id} FA)$, $\epsilon B = \theta_o\ (GB \xrightarrow{id} GB)$.

In the above situation, the following two diagrams commute:

(6)
$$\begin{array}{ccc} \mathbb{A}(A, A') & \xrightarrow{\mathbb{A}(A, \eta A')} & \mathbb{A}(A, GFA') \\ & \searrow^{F_{A,A'}} & \downarrow \theta_{A,FA'} \\ & & \mathbb{B}(FA, FA') \end{array}$$,

(7)
$$\begin{array}{ccc} \mathbb{B}(B, B') & \xrightarrow{\mathbb{B}(\epsilon B, B')} & \mathbb{B}(FGB, B') \\ & \searrow^{G_{B,B'}} & \downarrow \theta_{GB,B'} \\ & & \mathbb{A}(GB, GB') \end{array}$$,

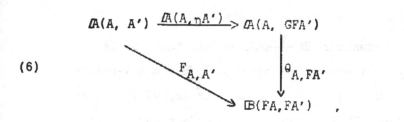

It is easy to verify the following:

<u>Proposition 0.1</u>

Let $\mathbb{A} \xrightarrow{F} \mathbb{B}$ be V-left adjoint to $\mathbb{B} \xrightarrow{G} \mathbb{A}$; then for any V-category $\mathbb{C}$ and V-functors $\mathbb{C} \xrightarrow{H} \mathbb{A}$, $\mathbb{C} \xrightarrow{H'} \mathbb{B}$, there is, naturally in H and H', a one to one and onto correspondence between V-natural transformations from H to GH' and V-natural transformations from FH to H'.

That is, given

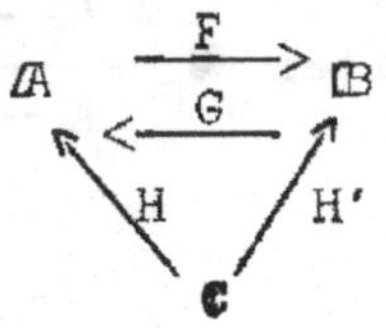

we have:

$$\frac{H \Longrightarrow GH'}{FH \Longrightarrow H'}$$ ∎

<u>Definition 0.3</u>

Given a V-functor $\mathbb{B} \xrightarrow{T} \mathbb{V}$, we say that T is <u>representable</u> if there is an object $B \in \mathbb{B}$ and a V-natural isomorphisms $\mathbb{B}(B, -) \overset{\theta}{\Longrightarrow} T$. The pair (B, θ) is called a <u>representation</u>, B the <u>representing object</u> and θ the <u>representing isomorphisms</u>.

The following (classical) very useful relation between adjoints and representable functors still holds in the V-context. Namely:

Proposition 0.2

Given a V-functor $\mathbb{B} \xrightarrow{G} \mathbb{A}$, G has a V-left adjoint $\mathbb{A} \xrightarrow{F} \mathbb{B}$ if and only if for every $A \in \mathbb{A}$ the V-functor $\mathbb{B} \xrightarrow{\mathbb{A}(A,G(-))} \mathbb{V}$ is representable.

Proof:

The nontrivial part of the statement is seen by letting FA be the representing object, the adjunction isomorphisms are the representing isomorphisms. Then η is gotten by (4) and the V-structure of F by (6). ∎

Remark 0.1

Suppose we are given a V-functor $\mathbb{B} \xrightarrow{G} \mathbb{A}$ such that for a fixed object $A \in \mathbb{A}$ the V-functor $\mathbb{B} \xrightarrow{\mathbb{A}(A,G(-))} \mathbb{V}$ is representable. Then, denoting the representing object by FA and the representing isomorphism by θ, formula (4) (for the fixed A) still makes sense, and the map $A \xrightarrow{\eta A} GFA$ so obtained still satisfies formula (2) for every $B \in \mathbb{B}$.

Finally, let us record the following proposition.

Proposition 0.3

Given any V-adjoint situation $\mathbb{B} \xrightarrow{G} \mathbb{A}$, $\mathbb{A} \xrightarrow{F} \mathbb{B}$, $FG \overset{\epsilon}{\Longrightarrow} id_{\mathbb{B}}$ $GF \overset{\eta}{\Longleftarrow} id_{\mathbb{A}}$ $(\epsilon, \eta): F \dashv_V G$.

Then:

a) F is V-full-and-faithful if and only if η is an isomorphism.

b) F is V-faithful if and only if ηA is a V-monomorphism for every $A \in \mathbb{A}$.

Dually:

a)* G is V-full-and-faithful if and only if ϵ is an isomorphism

b)* G is V-faithful if and only if ϵB is a V-epimorphism for every $B \in \mathbb{B}$.

The undefined terms above have the obvious definitions. For the proof take a hard look at diagrams (6), (7) above. ∎

CHAPTER I

COMPLETENESS CONCEPTS

Section 1. V-limits

In Set-based Category Theory limits are preserved by representable functors just by definition. For a general $\mathbb{V}$, there is no reason why this should be true, and, since this fact is the very essence of a limit, if we want to rescue for V-theory the essential results of ordinary category theory we are forced to consider as limits only those which are preserved by the representable functors. Hence the following definitions.

Definition I.1.1

Let $\mathbb{\Gamma} \xrightarrow{\Gamma} \mathbb{A}$ be a functor, where $\mathbb{\Gamma}$ is any category and $\mathbb{A}$ is a V-category. A cone $B \xrightarrow{p_\lambda} \Gamma\lambda$ over Γ is a V-limit of Γ if for every $A \in \mathbb{A}$, the cone

$\mathbb{A}(A, B) \xrightarrow{\mathbb{A}(A, p_\lambda)} \mathbb{A}(A, \Gamma\lambda)$ over $\mathbb{A}(A, \Gamma(-)) = \mathbb{A}(A,-)\circ\Gamma$

is a limit in $\mathbb{V}$. An immediate consequence of this definition is that (because the base functor preserves limits) V-limits are limits, and are precisely those limits which happen to be preserved by the functors $\mathbb{A}(A,-)$.

Recall that equivalent to the universal property of V-limits is the fact that there is a natural in A one to one and onto correspondence ι_o between the class of cones $A \xrightarrow{f_\lambda} \Gamma_\lambda$ and the hom set $\mathbb{A}_o(A, \text{V-}\varprojlim_\lambda \Gamma_\lambda)$.

We will write this in the form:

$$\frac{A \xrightarrow{f_\lambda} \Gamma_\lambda}{A \xrightarrow{f} \text{V-}\varprojlim_\lambda \Gamma_\lambda}\ \iota_o$$

Given a functor $\mathbb{\Gamma} \xrightarrow{\Gamma} \mathbb{A}$, $\mathbb{\Gamma}$ any category and $\mathbb{A}$ a V-category, we say that the V-limit of Γ exists if the limit of Γ exists and if it is a V-limit.

Given any other V-category $\mathbb{B}$ and a V-functor $\mathbb{B} \xrightarrow{G} \mathbb{A}$, if both $\text{V-}\varprojlim_\lambda \Gamma_\lambda$ and $\text{V-}\varprojlim_\lambda G\Gamma_\lambda$ exist, since V-limits are limits, there is a canonical map $G(\text{V-}\varprojlim_\lambda \Gamma_\lambda) \xrightarrow{z} \text{V-}\varprojlim_\lambda G\Gamma_\lambda$.

<u>Definition I.1.2</u>

A V-functor $\mathbb{B} \xrightarrow{G} \mathbb{A}$ preserves V-limits if for any functor $\mathbb{\Gamma} \xrightarrow{\Gamma} \mathbb{A}$, whenever $\text{V-}\varprojlim_\lambda \Gamma_\lambda$ exists, then $\text{V-}\varprojlim_\lambda G\Gamma_\lambda$ also exists and the canonical map is an isomorphism.

Recall that for any $V \in \mathbb{V}$, the functor $\mathbb{V}(V,-)$ preserves limits, and so in $\mathbb{V}$-itself V-limits and ordinary limits are the same. It follows then that the representable functors preserve V-limits.

Consider two functors $\mathbb{\Gamma} \overset{\Gamma}{\underset{\Gamma'}{\rightrightarrows}} \mathbb{A}$, $\mathbb{\Gamma}$ any category and $\mathbb{A}$ a V-category, and a natural transformation $\Gamma \xrightarrow{\varphi} \Gamma'$. If both $V\text{-}\varprojlim_{\lambda} \Gamma_\lambda$ and $V\text{-}\varprojlim_{\lambda} \Gamma'_\lambda$ exist, then there is a morphism $V\text{-}\varprojlim_{\lambda} \Gamma_\lambda \xrightarrow{V\text{-}\varprojlim \varphi_\lambda} V\text{-}\varprojlim_{\lambda} \Gamma'_\lambda$ making the diagrams:

$$\begin{array}{ccc} V\text{-}\varprojlim_{\lambda} \Gamma_\lambda & \xrightarrow{V\text{-}\varprojlim_{\lambda} \varphi_\lambda} & V\text{-}\varprojlim_{\lambda} \Gamma'_\lambda \\ \Big\downarrow p_\lambda & & \Big\downarrow p_\lambda \\ \Gamma_\lambda & \xrightarrow{\varphi_\lambda} & \Gamma'_\lambda \end{array}$$

commutative.

<u>Proposition I.1.1</u>

If φ_λ is a V-monomorphism for every λ, then so is $V\text{-}\varprojlim_{\lambda} \varphi_\lambda$. ∎

<u>Proposition I.1.2</u>

Given any V-category $\mathbb{A}$ and a V-meet diagram

$$\begin{array}{ccc} P & \longrightarrow & A \\ \downarrow & & \downarrow \\ B & \longrightarrow & C \end{array}$$

in $\mathbb{A}$, if the bottom row is a V-monomorphism then so is the top, if the right column is a V-monomorphism then so is the left. ∎

Finally, as in the ordinary case, given a functor $\mathbb{\Gamma} \times \mathbb{X} \xrightarrow{\Gamma} \mathbb{A}$ from a category $\mathbb{\Gamma} \times \mathbb{X}$ into a V-category $\mathbb{A}$, if for every λ $\text{V-}\underset{\overleftarrow{x}}{\lim}\, \Gamma(\lambda, x)$ exists, then it is a functor $\mathbb{\Gamma} \longrightarrow \mathbb{A}$. Similarly for $\text{V-}\underset{\overleftarrow{\lambda}}{\lim}\, \Gamma(\lambda,x)$. The following formula holds

$$(1) \quad \text{V-}\underset{\overleftarrow{\lambda}}{\lim}\, (\, \text{V-}\underset{\overleftarrow{x}}{\lim}\, \Gamma(\lambda,x)) \simeq \text{V-}\underset{\overleftarrow{x}}{\lim}\, (\, \text{V-}\underset{\overleftarrow{\lambda}}{\lim}\, \Gamma(\lambda,x))$$

If both inner V-limits exists, the two outer V-limits exist if and only if either one of them exists and they are equal.

<u>V-limits of V-functors</u>

Let $\mathbb{\Gamma}$ be any category, $\mathbb{A}$, $\mathbb{B}$ be V-categories and $\mathbb{\Gamma} \times \mathbb{B} \xrightarrow{\Gamma} \mathbb{A}$ a functor such that for every $\lambda \in \mathbb{\Gamma}$, $\mathbb{B} \xrightarrow{\Gamma(\lambda,-)} \mathbb{A}$ is a V-functor and for every $\lambda \xrightarrow{f} \mu \in \mathbb{\Gamma}$,

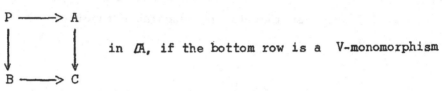

$\Gamma(\lambda,-) \Longrightarrow \Gamma(\mu,-)$ is a V-natural transformation.

Definition I.1.3

The V-limit of the V-functors $\Gamma(\lambda,-)$ is a V-functor $\mathbb{B} \longrightarrow \mathbb{A}$, denoted $\text{V-}\varprojlim_{\lambda} \Gamma(\lambda,-)$, and a cone of V-natural transformations
$\text{V-}\varprojlim_{\lambda} \Gamma(\lambda,-) \overset{p_\lambda}{\Longrightarrow} \Gamma(\lambda,-)$ satisfying the usual universal property with respect to cones of V-natural transformations. Equivalently, given any V-functor $\mathbb{B} \xrightarrow{F} \mathbb{A}$, there is, naturally in F, a one to one and onto correspondence ι_o between the class of V-natural transformations from F to $\text{V-}\varprojlim_{\lambda} \Gamma(\lambda,-)$ and the class of cones of V-natural transformations from F to the $\Gamma(\lambda,-)$'s. We will write this in the form:

$$\iota_o \; \frac{F \overset{\varphi_\lambda}{\Longrightarrow} \Gamma(\lambda,-)}{F \overset{\varphi}{\Longrightarrow} \text{V-}\varprojlim_{\lambda} \Gamma(\lambda,-)}$$

Proposition I.1.3

If for every $B \in \mathbb{B}$ the V-limit of the functor $\mathbb{C} \xrightarrow{\Gamma(-,B)} \mathbb{A}$ exists, then for every B, B' $\in \mathbb{B}$ there is a morphism:

$$\mathbb{B}(B, B') \longrightarrow \mathbb{A}(\text{V-}\varprojlim_{\lambda} \Gamma(\lambda,B), \ \text{V-}\varprojlim_{\lambda} \Gamma(\lambda,B'))$$

which gives to $\text{V-}\varprojlim_{\lambda} \Gamma(\lambda,B)$ the structure of a V-functor

$\mathbb{B} \longrightarrow \mathbb{A}$ in such a way that $\underset{\overleftarrow{\lambda}}{V\text{-lim}}\, \Gamma(\lambda, B) \xrightarrow{p_\lambda B} \Gamma(\lambda, B)$

are V-natural transformations (in B for every λ).

<u>Proof</u>:

Consider the diagram

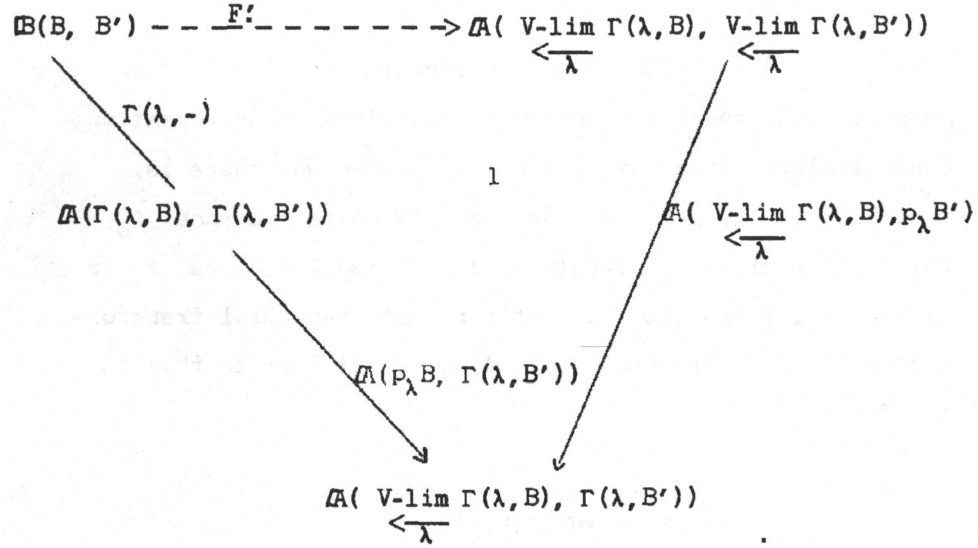

The right-hand diagonal is a limit because of the definition of V-limits. The existence of a unique morphism making the triangle commutative follows then from the fact that the left-hand diagonal is a cone, which we can see as follows:

Let $\lambda \xrightarrow{f} \mu$, then

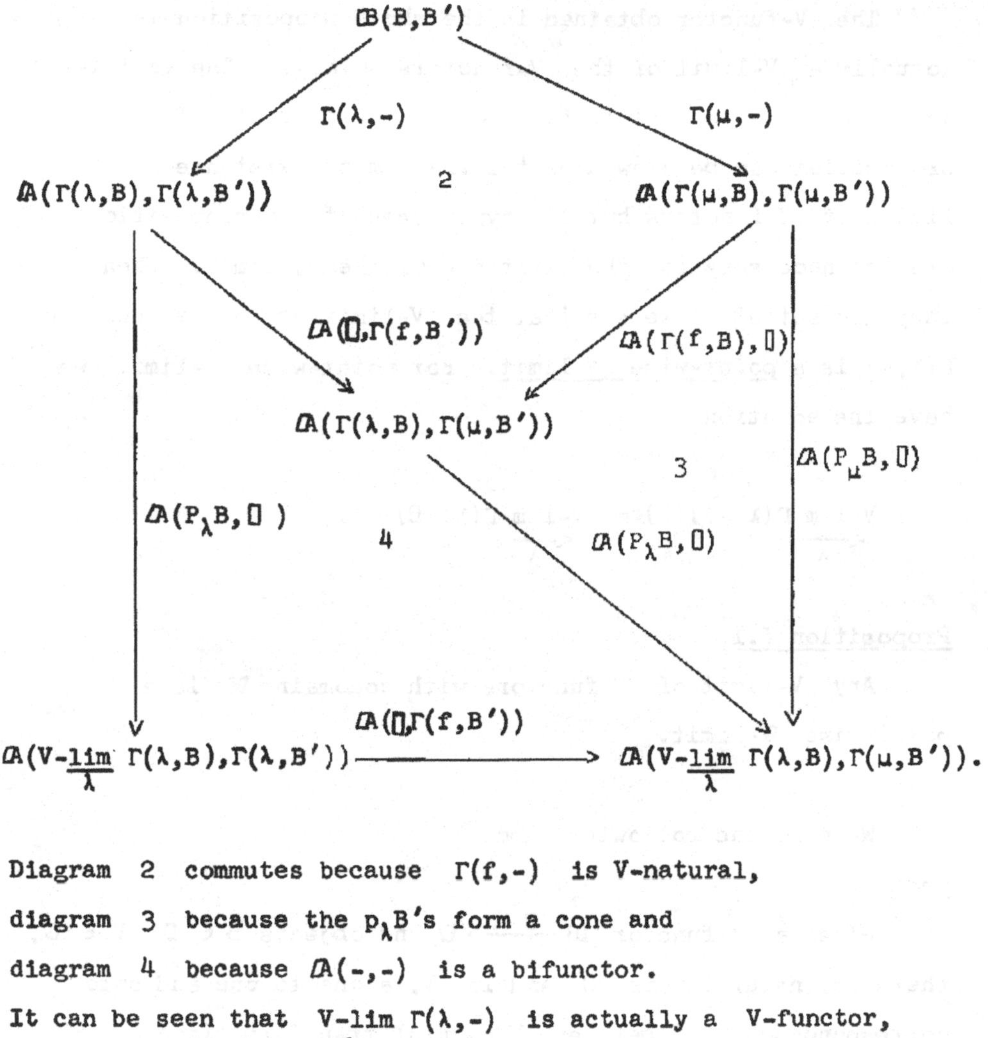

Diagram 2 commutes because $\Gamma(f,-)$ is V-natural, diagram 3 because the $p_\lambda B$'s form a cone and diagram 4 because $\mathbb{A}(-,-)$ is a bifunctor.

It can be seen that $V\text{-}\varprojlim_{\lambda} \Gamma(\lambda,-)$ is actually a V-functor,

and, finally, the commutativity of diagram 1 means exactly that p_λ is a V-natural transformation. ∎

The V-functor obtained in the above proposition is actually a V-limit of the V-functors $\Gamma(\lambda,-)$. The verification of this is straightforward but not trivial. The proposition can be viewed as the theorem of existence of V-limits of V-functors but the hypotheses of the proposition are not necessary for the existence of the V-limit. When they are satisfied we say that the V-limit of the V-functors $\Gamma(\lambda,-)$ is a <u>point-wise V-limit</u>. For point-wise V-limits we have the equation

$$(\underset{\overleftarrow{\lambda}}{\text{V-lim}}\, \Gamma(\lambda,-))(B) = \underset{\overleftarrow{\lambda}}{\text{V-lim}}\, \Gamma(\lambda, B) \quad .$$

<u>Proposition I.1.4</u>

Any V-limit of V-functors with codomain $\mathbb{V}$ is a point-wise $\mathbb{V}$-limit.

We need the following lemma:

<u>Lemma</u>:

Given a V-functor $\mathbb{B} \xrightarrow{G} \mathbb{V}$ and objects $B \in \mathbb{B}$, $V \in \mathbb{V}$, there is, naturally in G and in V, a one to one and onto correspondence e_o between V-natural transformations $\mathbb{B}(B, -) \otimes V \Longrightarrow G$ and morphisms $V \longrightarrow GB$. We write the above information as follows:

$$e_o \quad \frac{\mathbb{B}(B, -) \otimes V \Longrightarrow G}{V \longrightarrow GB}$$

Proof:

Using the fact that $\mathbb{V} \xrightarrow{-\otimes V} \mathbb{V}$ is V-left adjoint to $\mathbb{V} \xrightarrow{\mathbb{V}(V,-)} \mathbb{V}$ it follows from Proposition 0.1 that V-natural transformations $\mathbb{B}(B,-) \otimes V \Longrightarrow G$ correspond naturally in G with V-natural transformations $\mathbb{B}(B, -) \Longrightarrow \mathbb{V}(V, G(-))$ and it can be seen that this correspondence is also natural in V. Finally, by the representation theorem ([2]) these correspond with morphisms $V \longrightarrow GB$.

Proof of the Proposition:

In the situation (or data) of Definition I.1.3, with $\mathbb{A} = \mathbb{V}$, let B be any object of $\mathbb{B}$ and $V \xrightarrow{f_\lambda} \Gamma(\lambda, B)$ a cone over the functor $\Gamma \xrightarrow{\Gamma(-, B)} \mathbb{V}$. The following column of natural in V one to one and onto correspondences gives the desired result once we observe that cones correspond with cones under e_o because of its naturality in G.

$$V \xrightarrow{\ f\ } (\ \underset{\overleftarrow{\ \lambda\ }}{V\text{-lim}}\,\Gamma(\lambda,-))(B)$$

$$e_o \;\overline{\hspace{20em}}$$

$$\mathbb{B}(B,-) \otimes V \xRightarrow{\ \varphi\ } \underset{\overleftarrow{\ \lambda\ }}{V\text{-lim}}\,\Gamma(\lambda,-)$$

$$\ell_o \;\overline{\hspace{20em}}$$

$$\mathbb{B}(B,-) \otimes V \xRightarrow{\ \varphi_\lambda\ } \Gamma(\lambda,-)$$

$$e_o \;\overline{\hspace{20em}}$$

$$V \xrightarrow{\ f_\lambda\ } \Gamma(\lambda,B) \qquad \blacksquare$$

Given a V-functor $\mathbb{A} \xrightarrow{\ H\ } \mathbb{A}'$, there is a unique V-natural transformation z making the diagrams:

$$\begin{array}{ccc} H\,\underset{\overleftarrow{\ \lambda\ }}{V\text{-lim}}\,\Gamma(\lambda,-) & \xRightarrow{\ z\ } & \underset{\overleftarrow{\ \lambda\ }}{V\text{-lim}}\,H\Gamma(\lambda,-) \\ \Big\Downarrow H p_\lambda & \swarrow\!\!\!\swarrow\; p_\lambda & \\ H\Gamma(\lambda,-) & & \end{array}$$

commutative (when both V-limits exists).

Definition I.1.4

Suppose the V-limit of the V-functors $\Gamma(\lambda,-)$ exists, then we say that H preserves it if the V-limit of the V-functors $H\Gamma(\lambda,-)$ also exists and z is an isomorphism.

If a V-limit of V-functors is pointwise, it is clear that it is going to be preserved by all the representable functors.

However, this is not so for V-limits of V-functors in general. From Proposition I.1.4 we deduce:

<u>Proposition I.1.5</u>

A V-limit of V-functors is pointwise if and only if it is preserved by all the representables.

<u>Proof</u>: (in the notation of Definition I.1.3)

Let A and B be objects of $\mathbb{A}$, $\mathbb{B}$ respectively, and suppose the V-functor $\mathbb{A}(A,-)$ preserves the V-limit of the V-functors $\Gamma(\lambda,-)$.

Consider the commutative diagram:

$$\mathbb{A}(A,(\text{V-}\lim_{\overleftarrow{\lambda}} \Gamma(\lambda,-)(B)) \overset{zB}{\approx} (\text{V-}\lim_{\overleftarrow{\lambda}} \mathbb{A}(A,\Gamma(\lambda,\Gamma(\lambda,-)))(B)$$

$$\|$$

$$\text{V-}\lim_{\overleftarrow{\lambda}} \mathbb{A}(A,\Gamma(\lambda,B))$$

$$\mathbb{A}(A,p_\lambda B) \searrow \qquad \swarrow p_\lambda$$

$$\mathbb{A}(A,\Gamma(\lambda,B))$$

Then; precisely by definition of V-limits (Definition I.1.1) we have $(\text{V-}\lim_{\overleftarrow{\lambda}} \Gamma(\lambda,-))(B) = \text{V-}\lim_{\overleftarrow{\lambda}} \Gamma(\lambda,B)$ ∎

Similarly there is the dual concept of V-colimit and of V-colimit of V-functors.

Section 2. Tensors and Cotensors

Another basic feature of set-based category theory is the following: Suppose $\mathbb{A}$ is any category; A an object of $\mathbb{A}$ and S a set, and assume that the coproduct of S copies of A exists; then for every object $A' \in \mathbb{A}$ there is a bijection:

$$(1) \qquad \mathbb{A}(\coprod_S A, A') \approx \mathbb{S}(S, \mathbb{A}(A, A')).$$

Namely, this coproduct is a representing object for the functor $\mathbb{A} \xrightarrow{\mathbb{S}(S, \mathbb{A}(A, -))} \mathbb{S}$. We see then that, for example, the presence of coproducts guarantees the existence of left adjoints for the representable functors. This is an essential property of cocomplete categories which remains more or less hidden (in adjoint functor theorems, Kan Extensions, work on completions, etc.) because it is used in an implicit manner. Since formula (1) is no longer true in the V-case, and not even general colimits produce left adjoints for the representables, this fact should be established as an independent concept of the notions of cocompleteness, and has to be used explicitly in the V-version of the studies mentioned above.

Definition I.2.1$^{(op)}$

Let $\mathbb{A}$ be a V-category, $A \in \mathbb{A}$ an object of $\mathbb{A}$ and $V \in \mathbb{V}$ an object of $\mathbb{V}$. The tensor of V with A is a representation for the functor $\mathbb{A} \xrightarrow{\mathbb{V}(V,\, \mathbb{A}(A,\, -))} \mathbb{V}$, the representing object denoted $V \otimes_{\mathbb{A}} A$ and the representing isomorphism ω (in both directions). As in the case of colimits (colimit object and injections) we will often refer to the tensor just by its representing object.

The above representation explicitly takes the form:

$$\mathbb{A}(V \otimes_{\mathbb{A}} A,\, A') \overset{\omega}{\approx} \mathbb{V}(V,\, \mathbb{A}(A,\, A')) \,. \qquad \omega \circ \omega = id \,.$$

(for every $A' \in \mathbb{A}$, V-natural in A')

We will denote the bijection at the level of sets as:

$$\omega_o \; \frac{V \otimes_{\mathbb{A}} A \longrightarrow A'}{V \longrightarrow \mathbb{A}(A,\, A')}$$

Suppose that, for fixed A, the V-functors $\mathbb{V}(V,\, \mathbb{A}(A,\, -))$ are representable for every $V \in \mathbb{V}$. Then, by Proposition 0.2, $V \otimes_{\mathbb{A}} A$ becomes a V-functor, $\mathbb{V} \xrightarrow{-\otimes_{\mathbb{A}} A} \mathbb{A}$, V-left adjoint to $\mathbb{A} \xrightarrow{\mathbb{A}(A,-)} \mathbb{V}$.

Similarly, suppose that, for fixed V, the V-functors $\mathbb{V}(V,\, \mathbb{A}(A,-)$ are representable for every $A \in \mathbb{A}$. Then $V \otimes_{\mathbb{A}} A$ is a V-functor $\mathbb{A} \xrightarrow{V \otimes -} \mathbb{A}$, with a unique V-structure which renders ηV V-natural in A : (recall Remark 0.1).

$$V \xrightarrow{\eta V} \mathbb{A}(A, V \otimes_{\mathbb{A}} A), \; \eta V = \omega_o(V \otimes_{\mathbb{A}} A \xrightarrow{id} V \otimes_{\mathbb{A}} A)$$

$$(1) \quad (V \otimes_{\mathbb{A}} -)_{AA'} : \mathbb{A}(A, A') \xrightarrow{\mathbb{A}(-, V \otimes_{\mathbb{A}} A')}$$

$$\longrightarrow \mathbb{V}(\mathbb{A}(A', V \otimes_{\mathbb{A}} A'), \mathbb{A}(A, V \otimes_{\mathbb{A}} A')) \longrightarrow$$

$$\xrightarrow{\mathbb{V}(\eta V, \square)} \mathbb{V}(V, \mathbb{A}(A, V \otimes_{\mathbb{A}} A')) \longrightarrow$$

$$\xrightarrow{\omega} \mathbb{A}(V \otimes_{\mathbb{A}} A, V \otimes_{\mathbb{A}} A')$$

<u>Definition I.2.2</u>$^{(op)}$

A V-category $\mathbb{A}$ is said to be <u>Tensored</u> if all the representables $\mathbb{A} \xrightarrow{\mathbb{A}(A,-)} \mathbb{V}$ have a V-left adjoint, or equivalently, if for every $A \in \mathbb{A}$ and $V \in \mathbb{V}$, the tensor $V \otimes_{\mathbb{A}} A$ exists.

<u>Proposition I.2.1</u>$^{(op)}$

If A is tensored, $V \otimes_{\mathbb{A}} A$ is a V-bifunctor $\mathbb{V} \otimes \mathbb{A} \xrightarrow{- \otimes_{\mathbb{A}} -} \mathbb{A}$, and ω is V-natural in all the variables. ∎

Tensors are associative in the sense that there is, for V,V' in $\mathbb{V}$ and A in $\mathbb{A}$, a V-natural isomorphism

$(V \otimes V') \otimes_{\mathbb{A}} A \xrightarrow{\approx} V \otimes_{\mathbb{A}} (V' \otimes_{\mathbb{A}} A)$, and for the unit object $I \in \mathbb{V}$, $I \otimes_{\mathbb{A}} A \xrightarrow{\approx} A$. These isomorphisms are appropriately coherent.

The notion dual to that of tensor is called cotensor, and because most of the work in this paper is done on that side of the duality, it seems convenient to develop it explicitly.

<u>Definition I.2.1</u>

Let $\mathbb{A}$ be a V-category, $A \in \mathbb{A}$ an object of $\mathbb{A}$ and $V \in \mathbb{V}$ an object of $\mathbb{V}$. The <u>cotensor</u> of V with A is a representation for the functor $\mathbb{A}^{op} \xrightarrow{\mathbb{V}(V,\ \mathbb{A}(-,\ A))} \mathbb{V}$, the representing object is denoted by $\bar{\mathbb{A}}(V, A)$ and the representing isomorphism by σ (in both directions).

The above representation, explicitly, takes the form:

$$\mathbb{A}(A', \bar{\mathbb{A}}(V, A)) \overset{\sigma}{\approx} \mathbb{V}(V, \mathbb{A}(A' A)). \qquad \sigma \circ \sigma = \text{id}.$$

(For every $A' \in \mathbb{A}$, V-natural in A').

We denote the bijection at the level of sets by:

$$\sigma_o \quad \frac{A' \longrightarrow \bar{\mathbb{A}}(V, A)}{V \longrightarrow \mathbb{A}(A', A)} \quad .$$

With a fixed A; if the functors $\mathbb{V}(V, \mathbb{A}(- A))$ are representable for every V, by Proposition 0.2 we conclude that $\bar{\mathbb{A}}(V, A)$ becomes a V-functor $\mathbb{V}^{op} \xrightarrow{\bar{\mathbb{A}}(-, A)} \mathbb{A}$, V-adjoint on the right to $\mathbb{A}^{op} \xrightarrow{\mathbb{A}(- A)} \mathbb{V}$.

Definition I.2.2

A V-category $\mathbb{A}$ is said to be Cotensored if all the representables $\mathbb{A}^{op} \xrightarrow{\mathbb{A}(- A)} \mathbb{V}$ have a V-left adjoint; or equivalently. if for every $A \in \mathbb{A}$ and $V \in \mathbb{V}$, $\bar{\mathbb{A}}(V, A)$ exists.

Suppose that, for fixed V, $\bar{\mathbb{A}}(V, A)$ exists for every $A \in \mathbb{A}$, then $\bar{\mathbb{A}}(V, A)$ is a V-functor: $\mathbb{A} \xrightarrow{\bar{\mathbb{A}}(V, -)} \mathbb{A}$, with a V-structure which, of course, is gotten by the exact dualization of (1) (page 20).

Proposition I.2.1

If $\mathbb{A}$ is cotensored, $\bar{\mathbb{A}}(V, A)$ is a V-bifunctor $\mathbb{V}^{op} \otimes \mathbb{A} \xrightarrow{\bar{\mathbb{A}}(-, -)} \mathbb{A}$, and σ is V-natural in all variables. ∎

Note that just by definition we have the following formal identities: For $A \in \mathbb{A}$, $V \in \mathbb{V}$

$V \otimes_{\mathbb{A}} A = \overline{\mathbb{A}^{op}}(V, A)$ and $V \otimes_{\mathbb{A}^{op}} A = \bar{\mathbb{A}}(V, A)$

$\mathbb{A}$ is tensored if and only if $\mathbb{A}^{op}$ is cotensored and vice versa.

Note also that the base category $\mathbb{V}$ is always tensored and cotensored, with $\otimes_{\mathbb{V}} = \otimes$ and $\overline{\mathbb{V}}(-,-) = \mathbb{V}(-,-)$. Furthermore, the following diagram commutes:

$$
(1) \qquad
\begin{array}{ccc}
\mathbb{V}(U \otimes V,\ W) & \xrightarrow[\approx]{\mathbb{V}(c,\square)} & \mathbb{V}(V \otimes U,\ W) \\
\approx\big\downarrow \omega & & \approx\big\downarrow \omega \\
\mathbb{V}(U,\ \mathbb{V}(V,\ W)) & \xrightarrow[\approx]{\sigma} & \mathbb{V}(V,\ \mathbb{V}(U,\ W))
\end{array}
$$

where c is the symmetry.

A final observation is the following proposition:

<u>Proposition I.2.2</u>

Let $\mathbb{A}$ be any V-category and $V \in \mathbb{V}$ an object of $\mathbb{V}$, then, when they exist, the V-functor $\mathbb{A} \xrightarrow{V \otimes_{\mathbb{A}} -} \mathbb{A}$ is V-left adjoint to the V-functor $\mathbb{A} \xrightarrow{\overline{\mathbb{A}}(V,\ -)} \mathbb{A}$. The adjunction isomorphism is given by:

$$\mathbb{A}(V \otimes_{\mathbb{A}} A,\ B) \overset{\omega}{\approx} \mathbb{V}(V,\ \mathbb{A}(A,\ B)) \overset{\sigma}{\approx} \mathbb{A}(A,\ \overline{\mathbb{A}}(V,\ B)) \ . \ \blacksquare$$

Let $\mathbb{B} \xrightarrow{G} \mathbb{A}$ be a V-functor and let B, V be objects of $\mathbb{B}$ and $\mathbb{V}$ respectively. Suppose that both cotensors $\overline{\mathbb{B}}(V,\ B)$ and $\overline{\mathbb{A}}(V, GB)$ exist. Then there is a canonical morphism $G\,\overline{\mathbb{B}}(V,B) \xrightarrow{\ z\ } \overline{\mathbb{A}}(V,GB)$ which is gotten in the following way:

$$
(1)\qquad
\begin{array}{l}
\qquad\qquad G\overline{\mathbb{B}}(V,B) \xrightarrow{\ \ z\ \ } \overline{\mathbb{A}}(V,GB) \\
\sigma_0\ \overline{\hphantom{xxxxxxxxxxxxxxxxxxxxxxxxxxxxxxxxxxxxxxxxxxxxxxx}} \\
\qquad V \longrightarrow \mathbb{B}(\overline{\mathbb{B}}(V,B),B) \xrightarrow{\ \ G\ \ } \mathbb{A}(G\overline{\mathbb{B}}(V,B),GB) \\
\sigma_0\ \overline{\hphantom{xxxxxxxxxxxxxxxxxxxxxxxxxxx}} \\
\qquad \overline{\mathbb{B}}(V,B) \xrightarrow{\ \ id\ \ } \overline{\mathbb{B}}(V,B)
\end{array}
$$

By the representation theorem ([2]), z is the unique morphism making commutative the diagram:

$$
\begin{array}{ccc}
\mathbb{B}(-,\overline{\mathbb{B}}(V,B)) & \overset{\sigma}{\approx} & \mathbb{V}(V,\mathbb{B}(-,B)) \\
\Big\Downarrow G & & \\
\mathbb{A}(G(-),G(\overline{\mathbb{B}}(V,B)) & & \Big\Downarrow \mathbb{V}(\square,G) \\
\Big\Downarrow \mathbb{A}(\square,z) & & \\
\mathbb{A}(G(-),\overline{\mathbb{A}}(V,GB)) & \overset{\sigma}{\approx} & \mathbb{V}(V,\mathbb{A}(G(-),GB)) .
\end{array}
$$

Definition I.2.3

A V-functor $\mathbb{B} \xrightarrow{G} \mathbb{A}$ preserves cotensors if for any $B \in \mathbb{B}$, $V \in \mathbb{V}$, whenever $\bar{\mathbb{B}}(V, B)$ exists, then $\bar{\mathbb{A}}(V, GB)$ also exists and z is an isomorphism.

The representable functors always preserves cotensors; namely, we have the following proposition:

Proposition I.2.3

Given any V-category $\mathbb{A}$, and any object $B \in \mathbb{A}$; the functor $\mathbb{A} \xrightarrow{\mathbb{A}(B, -)} \mathbb{V}$ preserves cotensors. Furthermore, for any $A \in \mathbb{A}$ and $V \in \mathbb{V}$ such that $\bar{\mathbb{A}}(V, A)$ exists, the

maps $\mathbb{A}(B, \bar{\mathbb{A}}(V, A)) \overset{z}{\underset{\sigma}{\rightrightarrows}} \mathbb{V}(V, \mathbb{A}(B, A))$ are equal.

Proof:

It is clear (since σ is an isomorphism) that all we have to prove is the equation $z = \sigma$. By Remark 0.1 it follows that σ is the composite:

$$\mathbb{A}(B, \bar{\mathbb{A}}(V, A)) \xrightarrow{\mathbb{A}(-, A)} \mathbb{V}(\mathbb{A}(\bar{\mathbb{A}}(V, A), A), \mathbb{A}(B, A)) \longrightarrow$$

$$\xrightarrow{\mathbb{V}(\eta V, \square)} \mathbb{V}(V, \mathbb{A}(B, A)) \quad ,$$

where $\eta V = \sigma_0(\bar{\mathbb{A}}(V, A) \xrightarrow{id} \bar{\mathbb{A}}(V, A))$.

On the other hand; from the definition of z ((1) page 24 with G = A(B, -)) and the naturality of σ_0 it follows that z is the composite:

$$\mathbb{A}(B, \bar{\mathbb{A}}(V, A)) \xrightarrow{\sigma_0(\mathbb{A}(B, -))} \mathbb{V}(\mathbb{A}(\bar{\mathbb{A}}(V, A), A), \mathbb{A}(B, A)) \longrightarrow$$

$$\xrightarrow{\mathbb{V}(\eta V, \square)} \mathbb{V}(V, \mathbb{A}(B, A)) .$$

So all we need is to see that $\mathbb{A}(-, A)_{\bar{\mathbb{A}}(V, A), B}$ and $\mathbb{A}(B, -)_{\bar{\mathbb{A}}(V, A), A}$ correspond to each other under σ_0.

But, more generally, for any C ∈ $\mathbb{A}$ the following fact is true:

$$\sigma_0 \frac{\mathbb{A}(C, A) \xrightarrow{\mathbb{A}(B,-)_{C,A}} \mathbb{V}(\mathbb{A}(B, C), \mathbb{A}(B, A))}{\mathbb{A}(B, C) \xrightarrow{\mathbb{A}(-,A)_{C,B}} \mathbb{V}(\mathbb{A}(C, A), \mathbb{A}(B, A))} .$$

This follows directly from (1) page 23 and the definitions of the V-structure of the functors $\mathbb{A}(B, -)$ and $\mathbb{A}(-, A)$. (see [2]). ∎

Observe that we could have defined the cotensor of an object A ∈ $\mathbb{A}$ with an object V ∈ $\mathbb{V}$ as being an object $\bar{\mathbb{A}}(V, A) \in \mathbb{A}$ for which there is bijection

$$\sigma_0 \frac{A' \longrightarrow \bar{\mathbb{A}}(V, A)}{V \longrightarrow \mathbb{A}(A', A)}$$ natural in A'. Then, the definition of z ((1) page 24) still is possible, and therefore Definition I.2.3 still makes sense. From the previous

proposition we see then that the cotensors defined in Definition I.2.1 are precisely those cotensors as above which are preserved by the representables.

Cotensors of V-functors

Let $\mathbb{B} \xrightarrow{G} \mathbb{A}$ be a V-functor, $\mathbb{A}$, $\mathbb{B}$ any V-categories and $V \in \mathbb{V}$ an object of $\mathbb{V}$.

Definition I.2.4

The cotensor of the V-functor G with V is a V-functor $\mathbb{B} \longrightarrow \mathbb{A}$, denoted $\overline{\mathbb{A}^{\mathbb{B}}}(VG)$ (where $\mathbb{A}^{\mathbb{B}}$ is only a notational symbol) such that for every V-functor $\mathbb{B} \xrightarrow{F} \mathbb{A}$, there is naturally in F, a one to one and onto correspondence σ_o between the class of V-natural families $V \xrightarrow{fB} \mathbb{A}(FB, GB)$ and the class of V-natural transformations from F to $\overline{\mathbb{A}^{\mathbb{B}}}(V, G)$. As usual, we will write all the above information in the compact form:

$$\sigma_o \; \frac{V \xrightarrow{fB} \mathbb{A}(FB, GB)}{F \xRightarrow{f} \overline{\mathbb{A}^{\mathbb{B}}}(V, G)} \; .$$

Proposition I.2.4

If for every $B \in \mathbb{B}$ $\overline{\mathbb{A}}(V, GB)$ exists, then $\overline{\mathbb{A}}(V, GB)$ is a V-functor in B : $\mathbb{B} \longrightarrow \mathbb{A}$.

Proof:

If $\overline{\mathbb{A}}(V, A)$ exists for every $A \in \mathbb{A}$, then we have the composite $\mathbb{B} \xrightarrow{G} \mathbb{A} \xrightarrow{\overline{\mathbb{A}}(V, -)} \mathbb{A}$.

If this is not the case, the dual of formula (1)(page 20) for objects in $\mathbb{A}$ of the form GB, $B \in \mathbb{B}$ preceded by $\mathbb{B}(B\ B') \xrightarrow{G} \mathbb{A}(GB, GB')$ still gives the desired result. ∎

The V-functor so obtained is actually the cotensor of the V-functor G with V, but the hypotheses of the proposition are not necessary. When they are satisfied we say that the cotensor of G with V is a <u>pointwise cotensor</u>. For pointwise cotensors we have the formula:

$$\overline{\mathbb{A}^{\mathbb{B}}}(V, G)(B) = \overline{\mathbb{A}}(V, GB) \quad .$$

Given a V-functor $\mathbb{A} \xrightarrow{H} \mathbb{A}'$, if the cotensor of G with V and the cotensor of HG with V both exist, then there is a (canonical) V-natural transformation

$$H\ \overline{\mathbb{A}^{\mathbb{B}}}(V, G) \xRightarrow{z} \overline{\mathbb{A}'^{\mathbb{B}}}(V, HG).$$

Suppose the cotensor $\overline{\mathbb{A}^{\mathbb{B}}}(V, G)$ exists, then, we say that <u>H preserves it</u> if the cotensor $\overline{\mathbb{A}^{\mathbb{B}}}(V, HG)$ also exists and z is an isomorphism. Using the fact that the V-category $\mathbb{V}$ is cotensored, and hence every cotensor of a $\mathbb{V}$-functor into it is pointwise, it is easy to prove the following characterization of pointwise cotensors of $\mathbb{V}$-functors.

Proposition I.2.5

Given any V-functor $\mathbb{B} \xrightarrow{G} \mathbb{A}$ and an object $V \in \mathbb{V}$, the cotensor of G with V (if it exists) is pointwise if and only if it is preserved by the representables $\mathbb{A} \xrightarrow{\mathbb{A}(A, -)} \mathbb{V}$. ∎

Section 3 Ends

Another concept related to completeness which has arisen in the V-context is that of ends and coends. In set-based category theory this concept is just (or more properly, can be realized as) a particular kind of limit that, although notationally complicated in its greatest generality, has been convenient and successfully handled in all the practical cases by the use of comma categories as indexes. In the general V-case ends can be constructed by means of cotensors and V-limits but the use of comma categories (as it is known) is no longer possible. Finally, let us say that for V-categories lacking cotensors the concept seems to be completely independent.

Definition I.3.1

Given a V-category $\mathbb{C}$ and a V-bifunctor $\mathbb{C}^{op} \otimes \mathbb{C} \xrightarrow{T} \mathbb{V}$, the end of T is an object of $\mathbb{V}$, denoted $\int_C T(C, C)$, and a V-natural family of morphisms $\int_C T(C, C) \xrightarrow{p_C} T(C, C)$, one for each $C \in \mathbb{C}$, satisfying the following universal property: Given any other V-natural family $V \xrightarrow{fC} T(C,C)$ there exist a unique morphism $V \longrightarrow \int_C T(C, C)$ making the following diagrams commutative:

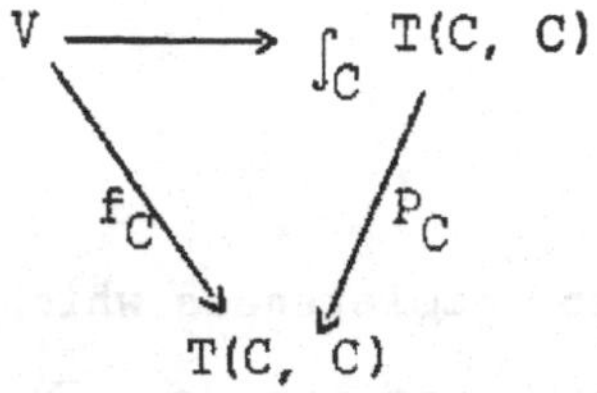

$C \in \mathbb{C}$.

<u>Definition I.3.2</u>

Given V-categories $\mathbb{C}$ and $\mathbb{A}$; and a V-bifunctor $\mathbb{C}^{op} \otimes \mathbb{C} \xrightarrow{T} \mathbb{A}$, the <u>end</u> of T is an object of $\mathbb{A}$, denoted $\int_C T(C, C)$, and a V-natural family of morphisms $\int_C T(C, C) \xrightarrow{pC} T(C, C)$, one for each $C \in \mathbb{C}$, such that for every object $A \in \mathbb{A}$; $\mathbb{A}(A, \int_C T(C, C)) \xrightarrow{\mathbb{A}(A, pC)} \mathbb{A}(A, T(C,C))$ is the end of $\mathbb{C}^{op} \otimes \mathbb{C} \xrightarrow{\mathbb{A}(A,T(-,-))} \mathbb{V}$.

Since giving a V-natural family $A \xrightarrow{fC} T(C, C)$ in $\mathbb{A}$ is the same as giving a V-natural family $I \xrightarrow{fC} \mathbb{A}(A, T(C, C))$ in $\mathbb{V}$, we see that the end of T satisfies the universal property of Definition I.3.1. Namely: given any V-natural family $A \xrightarrow{fC} T(C, C)$ there exists a unique morphism $A \longrightarrow \int_C T(C, C)$ making the following diagrams commutative:

$$\begin{array}{ccc} A & \longrightarrow & \int_C T(C,\ C) \\ & {\scriptstyle f_C}\searrow \quad \swarrow{\scriptstyle p_C} & \\ & T(C,\ C) & \end{array}$$

As usual, we call the morphisms p_C the projections, and we will often refer to the end as just the object $\int_C T(C,\ C)$.

Equivalent to this universal property is the fact that there is, naturally in A, a one to one and onto correspondence e_o between the class of V-natural families $A \xrightarrow{fC} T(C,\ C)$ and the hom set $\mathbb{A}_o\ (A,\ \int_C T(C,\ C))$, that, as usual, we denote:

$$e_o \ \frac{A \xrightarrow{fC} T(C,\ C)}{A \xrightarrow{F} \int_C T(C,\ C)} \quad .$$

Observe that this property alone is not enough, but that also preservation by the representables is required.

Let $\mathbb{C}^{op} \otimes \mathbb{C} \xrightarrow{T} \mathbb{B}$ be a V-bifunctor and $\mathbb{B} \xrightarrow{G} \mathbb{A}$ a V-functor. Suppose both the ends $\int_C T(C,\ C)$ and $\int_C GT(C,\ C)$ exist. Then there is a canonical (unique) morphism $G\int_C T(C,\ C) \xrightarrow{z} \int_C GT(C,\ C)$ making the following diagrams commutative:

$$\begin{array}{ccc} G \int_C T(C, C) & \xrightarrow{z} & \int_C GT(C, C) \\ & \searrow^{F(p_C)} & \downarrow{p_C} \\ & & FT(C, C) \end{array}$$

<u>Definition I.3.3</u>

A V-functor $\mathbb{B} \xrightarrow{G} \mathbb{A}$ <u>preserves</u> <u>ends</u> if for any V-bifunctor $\mathbb{C}^{op} \otimes \mathbb{C} \xrightarrow{T} \mathbb{B}$, whenever $\int_C T(C, C)$ exists, then $\int_C GT(C, C)$ also exists and the canonical map $G \int_C T(C, C) \xrightarrow{z} \int_C GT(C, C)$ is an isomorphism.

Let us remark that just by definition we have the formula:

$\mathbb{A}(A, \int_C T(C, C)) \simeq \int_C \mathbb{A}(A, T(C, C))$ meaning that the representable functors preserve ends.

Given two V-bifunctors $\mathbb{C}^{op} \otimes \mathbb{C} \overset{T}{\underset{T'}{\rightrightarrows}} \mathbb{A}$ and a V-natural transformation $T \overset{\varphi}{\Longrightarrow} T'$. If both ends exist, there is a unique morphism

(1) $\int_C T(C, C) \xrightarrow{\int_C \varphi(C, C)} \int_C T'(C, C)$ making the

diagrams

$$\begin{array}{ccc} \int_C T(C, C) & \xrightarrow{\int_C \varphi(C, C)} & \int_C T'(C, C) \\ \downarrow{\scriptstyle p_C} & & \downarrow{\scriptstyle p_C} \\ T(C, C) & \xrightarrow{\varphi(C, C)} & T'(C, C) \end{array}$$

commutative.

As in the case of V-limits, we have:

Proposition I.3.1

If $\varphi(C, C)$ is a V-monomorphism for every $C \in \mathbb{C}$, then so is $\int_C \varphi(C, C)$. ∎

Ends of V-functors

Let $\mathbb{C}$, $\mathbb{A}$, and $\mathbb{B}$ be any V-categories and $\mathbb{C}^{op} \otimes \mathbb{C} \otimes \mathbb{B} \xrightarrow{T} \mathbb{A}$ a V-functor. Then for each $C \in \mathbb{C}$ we have a V-functor $\mathbb{B} \xrightarrow{T(C,C,-)} \mathbb{A}$. Given a V-functor $\mathbb{B} \xrightarrow{F} \mathbb{A}$, by a V-family of V-natural transformations from F to the $T(C, C, -)$'s we understand a family, indexed by C, of V-natural transformations $F \xrightarrow{\theta C} T(C, C, -)$ such that for every $B \in \mathbb{B}$; the family of morphisms $FB \xrightarrow{\theta CB} T(C, C, B)$ is V-natural in C.

Definition I.3.4

The end of the V-functors $\mathbb{B} \xrightarrow{T(C, C, -)} \mathbb{A}$ is a V-functor $\mathbb{B} \longrightarrow \mathbb{A}$, denoted by $\int_C T(C, C, -)$ and a V-family of V-natural transformations

$\int_C T(C, C, -) \xRightarrow{pC} T(C, C, -)$ satisfying the universal property: given any other V-functor $\mathbb{B} \xrightarrow{F} \mathbb{A}$ together with a V-family of V-natural transformations $F \xRightarrow{\theta C} T(C, C, -)$, there is a unique V-natural transformation $F \Longrightarrow \int_C T(C, C, -)$ such that the diagram

$$\begin{array}{ccc} F & \Longrightarrow & \int_C T(C, C, -) \\ & {\scriptstyle \theta C}\searrow\!\!\!\searrow \quad \swarrow\!\!\!\swarrow{\scriptstyle pC} & \\ & T(C, C, -) & \end{array}$$

commutes.

Equivalently, given any V-functor $\mathbb{B} \xrightarrow{F} \mathbb{A}$, there is, naturally in F, a one to one and onto correspondence e_o between the class of V-natural transformations from F to $\int_C T(C, C, -)$ and the class of V-families of V-natural transformations from F to the T(C, C, -)'s. We denote this correspondence:

$$e_o \quad \frac{F \xRightarrow{\theta C} T(C, C, -)}{F \xRightarrow{\theta} \int_C T(C, C, -)} \quad .$$

<u>Proposition I.3.2</u>

If for every $B \in \mathbb{B}$ the end of the V-bifunctor $\mathbb{C}^{op} \otimes \mathbb{C} \xrightarrow{T(-,\,-,\,B)} \mathbb{A}$ exists, then for every $B, B' \in \mathbb{B}$ there is a unique morphism

$$\mathbb{B}(B, B') \longrightarrow \mathbb{A}\left(\int_C T(C, C, B), \int_C T(C, C, B')\right)$$

which gives to $\int_C T(C, C, -)$ the structure of a V-functor $\mathbb{B} \longrightarrow \mathbb{A}$ in such a way that for every $C \in \mathbb{C}$

$\int_C T(C, C, B) \xrightarrow{p_C{}^B} T(C, C, B)$ is a V-natural

transformation in B .

<u>Proof</u>:

$$\begin{array}{ccc}
\mathbb{B}(B, B') & \dashrightarrow & \mathbb{A}\left(\int_C T(C,C,B), \int_C T(C,C,B')\right) \\
\Big\downarrow {\scriptstyle T(C,\,C,\,-)} & & \Big\downarrow {\scriptstyle \mathbb{A}(\int_C T(C,C,B),\, p_C{}^{B'})} \\
\mathbb{A}(T(C, C, B), T(C, C, B')) & \xrightarrow{\mathbb{A}(p_C{}^B,\; T(C,C,B'))} & \mathbb{A}\left(\int_C T(C,C,B), T(C,C,B')\right)
\end{array}$$

The right column is an end just by definition (Definition I.3.2) The existence of a unique morphism making the diagram commutative follows then from the fact that the left column and bottom arrows are V-natural in C. It can be seen that this is actually a V-functor structure, and, finally, the commutativity of the square means exactly that p_C is a V-natural transformation. ∎

The V-functor obtained in the above proposition is actually an end of the V-functors $T(C, C, -)$. The proposition can be viewed as a theorem of existence of ends of V-functors, and, as in the case of V-limits, the hypotheses are not necessary. When they are satisfied we say that the end is a pointwise end. For pointwise ends we have the equation

$$(\int_C T(C, C, -))(B) = \int_C T(C, C, B) \quad .$$

Given a V-functor $\mathbb{A} \xrightarrow{H} \mathbb{A}'$; paraphrasing the definition of the case of V-limits (Definition I.1.4) we define preservation of an end of V-functors by H and we have the characterization of pointwise ends:

Proposition I.3.1

An end of V-functors is pointwise if and only if it is preserved by the representables. ∎

Suppose that in the situation (or data) of Definition I.3.4 $\mathbb{B}$ is of the form $\mathbb{B} = \mathbb{D} \otimes \mathbb{D}^{op}$, and hence, T is a V-functor $\mathbb{C}^{op} \otimes \mathbb{C} \otimes \mathbb{D}^{op} \otimes \mathbb{D} \xrightarrow{T} \mathbb{A}$. Then, the following formula holds:

$$(1) \quad \int_C \int_D T(C, C, D, D) \approx \int_D \int_C T(C, C, D, D) \quad .$$

If both inner ends exist, and they are pointwise, the two outer ends exist if and only if either one of them exist and they are equal.

The above formula follows trivially from a more general proposition that we will need later:

<u>Proposition I.3.4</u>

If the end of V-functors $\int_D T(-, -, D, D)$ exist and it is pointwise; then there is a one to one and onto correspondence between V-natural families $A \xrightarrow{fCD} T(C, C, D, D)$ and V-natural families $A \xrightarrow{fC} \int_D T(C, C, D, D)$. ∎

Suppose now that the V-category B is replaced by any category $\mathbb{\Gamma}$. Explicitly; let $\mathbb{C}^{op} \otimes \mathbb{C} \times \mathbb{\Gamma} \xrightarrow{T} \mathbb{A}$ a functor such that for every $\lambda \in \mathbb{\Gamma}$ $\mathbb{C}^{op} \otimes \mathbb{C} \xrightarrow{T(-, -, \lambda)} \mathbb{A}$ is a V-bifunctor and for every $\lambda \xrightarrow{f} \mu \in \mathbb{\Gamma}$, $T(-, -, \lambda) \Longrightarrow T(-, -, \mu)$ is a V-natural transformation. If for every $\lambda \in \mathbb{\Gamma}$ the end $\int_C T(C, C, \lambda)$ exists then (using (1) page 32) $\int_C T(C, C, -)$ is a functor $\mathbb{\Gamma} \longrightarrow \mathbb{A}$ and $\int_C T(C, C, \lambda) \xrightarrow{pC\lambda} T(C, C, \lambda)$ are natural transformations (in λ for every C). Suppose the V-limit of the V-functor $\mathbb{C}^{op} \otimes \mathbb{C} \xrightarrow{T(-, -, \lambda)} \mathbb{A}$ exists and it is pointwise. Then the following formula holds:

$$(1) \qquad \underset{\overleftarrow{\lambda}}{\text{V-lim}} \int_C T(C, C, \lambda) \approx \int_C \underset{\overleftarrow{\lambda}}{\text{V-lim}} \; T(C, C, \lambda)$$

If both the inner end and the inner V-limit exist, then the outer end exists if and only if the outer V-limit exist, and they are equal.

<u>Construction of ends</u>

Given a V-bifunctor $\mathbb{C}^{op} \otimes \mathbb{C} \xrightarrow{T} \mathcal{A}$, the usual criteria for the V-naturality of a family $A \xrightarrow{f_C} T(C, C)$ is the commutativity of the following diagram:

$$\begin{array}{ccc}
\mathbb{C}(C, C') & \xrightarrow{\quad T(C,\, -) \quad} & \mathcal{A}(T(C,C),\, T(C,C')) \\
\Big\downarrow{\scriptstyle T(-,\, C')} & & \Big\downarrow{\scriptstyle \mathcal{A}(f_C,\, \square)} \\
\mathcal{A}(T(C', C'),\, T(C,C')) & \xrightarrow{\quad \mathcal{A}(f_{C'},\, \square) \quad} & \mathcal{A}(A,\, T(C, C'))
\end{array}$$

which, if $\mathcal{A}$ has cotensors, by naturality of σ_o is equivalent to the commutativity of :

$$\begin{array}{ccc}
A & \xrightarrow{\quad fC \quad} & T(C, C) \\
\Big\downarrow{\scriptstyle f_{C'}} & & \Big\downarrow{\scriptstyle \sigma_o(T(C, -))} \\
T(C', C') & \xrightarrow{\ \sigma_o(T(-, C'))\ } & \mathbb{A}(\mathbb{C}(C, C'), T(C, C'))
\end{array}$$

from which it readily follows that the end of T is actually the V-limit of a diagram in $\mathbb{A}$.

<u>Proposition I.3.5</u>

Given a bifunctor $\mathbb{C}^{op} \otimes \mathbb{C} \xrightarrow{T} \mathbb{A}$, where $\mathbb{A}$ is a cotensored V-category, the end of T, if it exists, is the V-limit of a diagram constructed by the aid of cotensors. ∎

The notion dual to that of end is called <u>coend</u>, we denote it by $T(C, C) \xrightarrow{\lambda C} \int^C T(C, C)$, we refer to the morphisms λ_C as the <u>injections</u> and, as usual, to the object $\int^C T(C, C)$ as the coend.

<u>Section 4 Kan Extensions</u>

<u>Definition I.4.1</u>

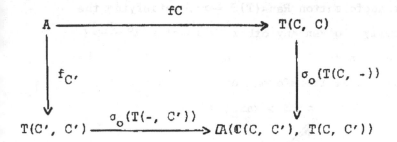

Given a V-functor $\mathbb{C} \xrightarrow{S} \mathbb{B}$, a V-category $\mathbb{A}$ and a V-functor $\mathbb{C} \xrightarrow{T} \mathbb{A}$, <u>the Right Kan extension of T along S</u> is a V-functor $\mathbb{B} \longrightarrow \mathbb{A}$, denoted $\mathrm{Ran}_S(T)$, and a V-natural transformation $\mathrm{Ran}_S(T)S \overset{\epsilon}{\Longrightarrow} T$, satisfying the universal property: given any other V-functor $\mathbb{B} \xrightarrow{F} \mathbb{A}$ together with a V-natural transformation $FS \overset{\varphi}{\Longrightarrow} T$, there is a unique V-natural transformation $F \overset{\psi}{\Longrightarrow} \mathrm{Ran}_S(T)$ such that the diagram

$$\begin{array}{ccc} FS & \overset{\psi S}{\Longrightarrow} & \mathrm{Ran}_S(T)S \\ & {\scriptstyle\varphi}\searrow\!\!\searrow \quad \swarrow\!\!\swarrow{\scriptstyle\epsilon} & \\ & T & \end{array}$$

commutes.

We have the configuration

$$\begin{array}{ccc} \mathbb{C} & \xrightarrow{S} & \mathbb{B} \\ & {\scriptstyle T}\searrow \overset{\epsilon}{\Longleftarrow} \swarrow{\scriptstyle \mathrm{Ran}_S(T)} & \\ & \mathbb{A} & \end{array}$$

For any V-functor $\mathbb{B} \xrightarrow{F} \mathbb{A}$, there is, naturally in F, a one to one and onto correspondence r_o between the class of V-natural transformations from FS to T and the class of V-natural transformations from F to $\mathrm{Ran}_S(T)$.

$r_o(F \overset{\psi}{\Longrightarrow} \mathrm{Ran}_S(T)) = (FS \overset{\psi S}{\Longrightarrow} \mathrm{Ran}_S(T)S \overset{\epsilon}{\Longrightarrow} T)$.

As usual, we will denote this correspondence:

$$r_o \; \frac{FS \Longrightarrow T}{F \Longrightarrow \mathrm{Ran}_S(T)}$$

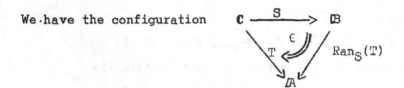

We can recover ϵ from r_o by means of $\epsilon = r_o\ (\mathrm{Ran}_S(T) \xrightarrow{id} \mathrm{Ran}_S(T))$ and both sets of data are entirely equivalent, the fact that r_o is one to one and onto means the universal property of ϵ and vice versa.

The (meta) adjoint situation underneath this definition guarantees that, in case of existence of the extension involved, $\mathrm{Ran}_S(T)$ behaves "functorially" in T in a way such that r_o becomes natural. Given any $T \xRightarrow{\varphi} T'$; then the diagram:

$$\begin{array}{ccc} \mathrm{Ran}_S(T)S & \xRightarrow{\mathrm{Ran}_S(\varphi)S} & \mathrm{Ran}_S(T')S \\ \Downarrow \epsilon & & \Downarrow \epsilon \\ T & \xRightarrow{\varphi} & T' \end{array}$$

commutes.

Also, if $T \xRightarrow{\varphi} T'$ is a V-natural isomorphism, then $\mathrm{Ran}_S(T)$ exists if and only if $\mathrm{Ran}_S(T')$ does.

Let $\begin{array}{ccc} \mathbb{C} & \xrightarrow{S} & \mathbb{B} \\ & \searrow^{T} & \\ & & \mathbb{A} \end{array}$ be two V-functors and

$\mathbb{A} \xrightarrow{H} \mathbb{A}'$ a V-functor. Suppose both the right kan extensions $\mathrm{Ran}_S(T)$ and $\mathrm{Ran}_S(HT)$ exist. Then there is a

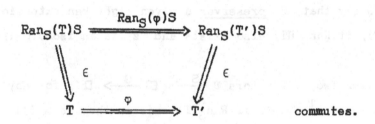

canonical (unique) V-natural transformation
$H\ \mathrm{Ran}_S(T) \overset{z}{\Longrightarrow} \mathrm{Ran}_S(HT)$ making the following diagram commutative:

$$\begin{array}{ccc} H\ \mathrm{Ran}_S(T)S & \overset{zS}{\Longrightarrow} & \mathrm{Ran}_S(HT)S \\ & {}_{H\epsilon}\searrow \quad \swarrow_{\epsilon} & \\ & HT & \end{array}$$

(1) $z = r_o(H\epsilon)$

<u>Definition I.4.2</u>

We say that H <u>preserves</u> a given right kan extension $\mathrm{Ran}_S(T)$, if $\mathrm{Ran}_S(HT)$ also exists and z is an isomorphism.

Given two V-functors $\mathbb{C} \xrightarrow{S} \mathbb{B} \xrightarrow{G} \mathbb{B}'$, for any V-functor $\mathbb{C} \xrightarrow{T} \mathbb{A}$, if $\mathrm{Ran}_S(T)$ exists, then $\mathrm{Ran}_{GS}(T)$ exists if and only if $\mathrm{Ran}_G\ (\mathrm{Ran}_S(T))$ exists and they are equal.

$$\mathrm{Ran}_{GS}(T) = \mathrm{Ran}_G(\mathrm{Ran}_S(T)) \quad .$$

To prove this, consider the two columns:

$$\begin{array}{ll} & FGS \Longrightarrow T \\ r_o & \overline{\qquad\qquad\qquad} \\ & FG \Longrightarrow \mathrm{Ran}_S(T) \\ r_o & \overline{\qquad\qquad\qquad} \\ & F \Longrightarrow \mathrm{Ran}_G(\mathrm{Ran}_S(T)) \end{array} \qquad\qquad \begin{array}{ll} & FG \Longrightarrow \mathrm{Ran}_S(T) \\ r_o & \overline{\qquad\qquad\qquad} \\ & FGS \Longrightarrow T \\ r_o & \overline{\qquad\qquad\qquad} \\ & F \Longrightarrow \mathrm{Ran}_{GS}(T) \end{array}$$

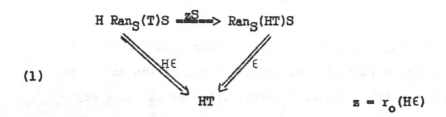

Assuming the existence of $Ran_S(T)$, the left column proves that $Ran_G(Ran_S(T))$ satisfies the definition of $Ran_{GS}(T)$, while the right column proves that $Ran_{GS}(T)$ satisfies the definition of $Ran_G(Ran_S(T))$.

Of course, if we have already the three extensions to begin with, then the above equality becomes a V-natural isomorphism, θ appropriately coherent with respect to the three r_o's (the two columns above show the equations which are satisfied). Also, the following diagram commutes:

(1)

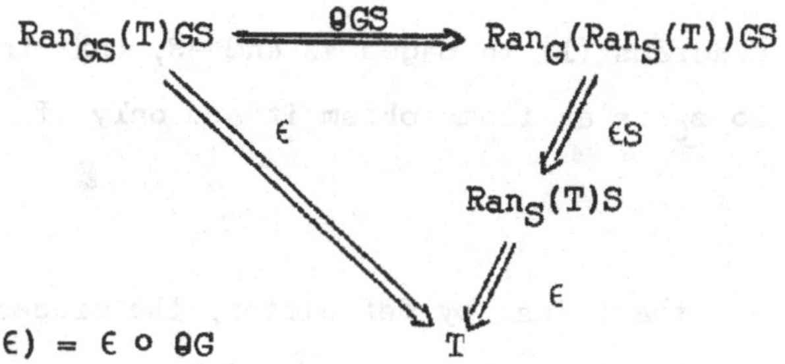

$r_o(\epsilon) = \epsilon \circ \theta G$

Now, assume that $Ran_S(T)$ exists. Then, for any V-functor $\mathbb{A} \xrightarrow{H} \mathbb{A}'$ which preserves it, the following statement holds:

<u>Proposition I.4.1</u>

In the above situation, H preserves $Ran_{GS}(T)$ if and only if it preserves $Ran_G(Ran_S(T))$ (if either of the two (hence both) exists) .

<u>Proof</u>:

Consider the diagram.

$$\begin{array}{ccccc}
H\, Ran_G(Ran_S(T)) & \overset{H\theta}{\approx} & H(Ran_{GS}(T)) & & \\
\Big\Downarrow z_0 & & & \searrow\!\!\searrow z_1 & \\
Ran_G(HRan_S(T)) & \overset{Ran_G(z)}{\approx} & Ran_G(Ran_S(HT)) & \overset{\theta}{\approx} & Ran_{GS}(HT)
\end{array}$$

The existence of the extensions not assumed in the hypothesis follows from considerations made before. It can be seen, from diagrams (1) in pages 42 and 43, that the diagram commutes, so z_0 is an isomorphism if and only if z_1 is. ∎

Besides the fact that, just by definition, the process of taking Right Kan extensions along a fixed functor $\mathbb{C} \xrightarrow{S} \mathbb{B}$ provides a (meta) right adjoint to the process of composing with S on the right, there is another much more intimate and important connection which relates <u>single</u> Kan extensions and <u>legitimate</u> adjoints.

First let us observe that, unlike the representables, V-functors having V-left adjoint satisfy the strong continuity property:

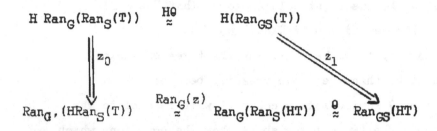

<u>Proposition I.4.2</u>

If a functor has a V-left adjoint, then it preserves any right-kan extensions that might exist.

<u>Proof</u>:

Let

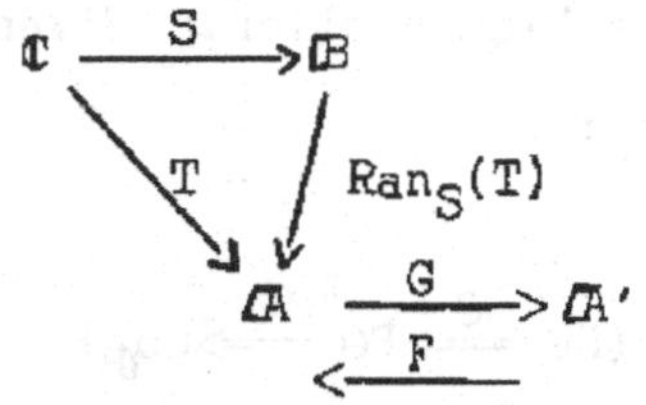

$F \dashv_V G$.

then, given any V-functor $\mathbb{B} \xrightarrow{H} \mathbb{A}$, consider:

$$\begin{array}{c} H \Longrightarrow G\,Ran_S(T) \\ \hline FH \Longrightarrow Ran_S(T) \\ r_0 \;\overline{} \\ FHS \Longrightarrow T \\ \hline HS \Longrightarrow GT\ . \end{array}$$

The two unlabeled passages are provided by Proposition 0.1. So $G\,Ran_S(T)$ satisfies the definition of $Ran_S(GT)$ ∎

<u>Theorem I.4.1</u> (Formal criteria of existence of adjoint)

Given any V-functor $\mathbb{B} \xrightarrow{G} \mathbb{A}$, G has a V-left adjoint $\mathbb{A} \xrightarrow{F} \mathbb{B}$ if and only if $\mathbb{A} \xrightarrow{Ran_G(id_{\mathbb{B}})} \mathbb{B}$ exists and is preserved by G. Moreover:

$$F = \mathrm{Ran}_G(id_{\mathbb{B}}) \ .$$

<u>Proof</u>:

Suppose G has a V-left adjoint F. Let $FG \overset{\epsilon}{\Longrightarrow} id_{\mathbb{B}}$ and $id_{\mathbb{A}} \overset{\eta}{\Longrightarrow} GF$ be a V-adjunction. Given any V-functor $\mathbb{A} \xrightarrow{H} \mathbb{B}$ define r_o:

$$r_o \ \frac{H \Longrightarrow F}{HG \Longrightarrow id_{\mathbb{B}}} \qquad \begin{array}{l} r_o(H \overset{\varphi}{\Longrightarrow} F) = (HG \overset{\varphi G}{\Longrightarrow} FG \overset{\epsilon}{\Longrightarrow} id_{\mathbb{B}}) \\ r_o(HG \overset{\psi}{\Longrightarrow} id_{\mathbb{B}}) = (H \overset{H\eta}{\Longrightarrow} HGF \overset{\psi F}{\Longrightarrow} F) \ . \end{array}$$

Since ϵ and η are V-natural, r_o sends V-natural transformations into V-natural transformations. That the two composites $r_o \circ r_o$ equal the identity follows easily from the triangular equations and naturality. So r_o is a one to one and onto correspondence. Finally, the naturality of r_o in H offers no difficulty. Conversely, suppose $\mathrm{Ran}_G(id_{\mathbb{B}})$ exists and $G\,\mathrm{Ran}_G(id_{\mathbb{B}}) \overset{z}{\Longrightarrow} \mathrm{Ran}_G(G)$ is an isomorphism. Then, there is a V-adjunction $(\epsilon,\eta): \mathrm{Ran}_G(id_{\mathbb{B}}) \longrightarrow\!\!|_V G$ defined as follows:

$$r_o \ \frac{G \overset{id}{\Longrightarrow} G}{id_{\mathbb{A}} \overset{\eta'}{\Longrightarrow} \mathrm{Ran}_G(G) \overset{z}{\Longleftarrow} G\,\mathrm{Ran}_G(id_{\mathbb{B}})} \qquad \eta = z^{-1} \circ \eta'$$

and $r_o \dfrac{Ran_G(id) \overset{id}{\Longrightarrow} Ran_G(id)}{Ran_G(id)\, G \overset{\epsilon}{\Longrightarrow} id}$.

Recall that we also have $r_o \dfrac{Ran_G(G) \overset{id}{\Longrightarrow} Ran_G(G)}{Ran_G(G) G \overset{\epsilon'}{\Longrightarrow} G}$.

The triangular equations are:

a) 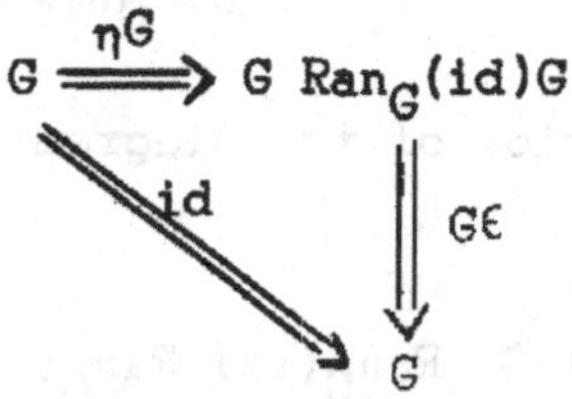

b) 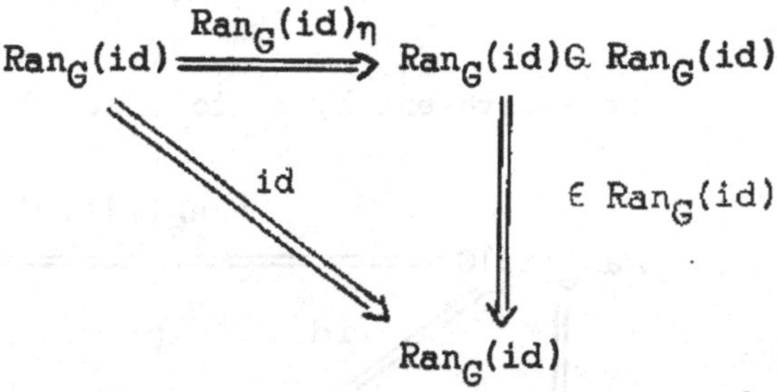

.

We prove them as follows:

a) 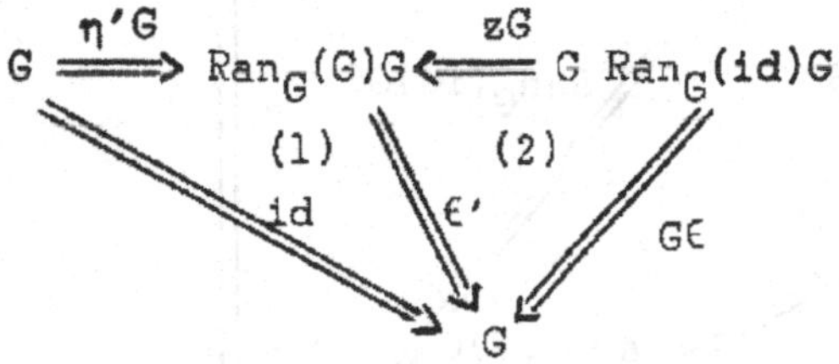

Diagram (2) commutes because of (1) (page 42).

The commutativity of diagram (1) is equivalent by r_o to that

of id 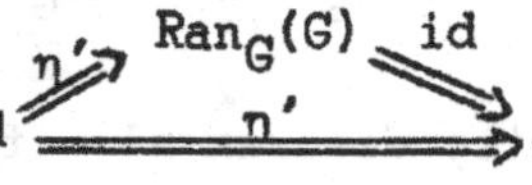

$Ran_G(G)$, so diagram (1) commutes.

b) The commutativity of the diagram:

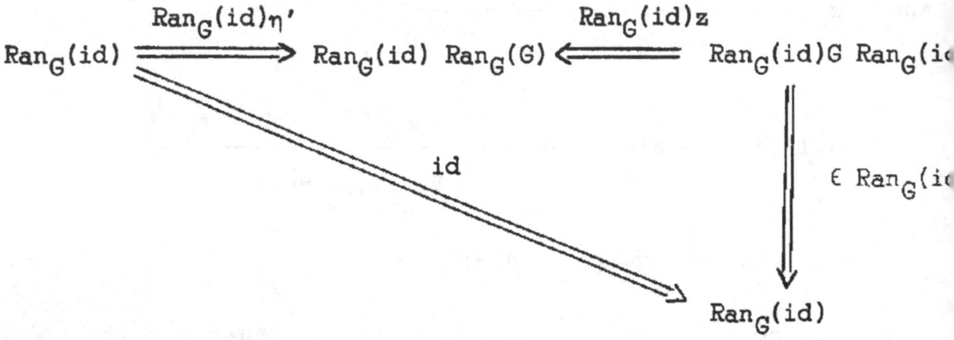

is equivalent by r_o to that of the exterior of the diagram:

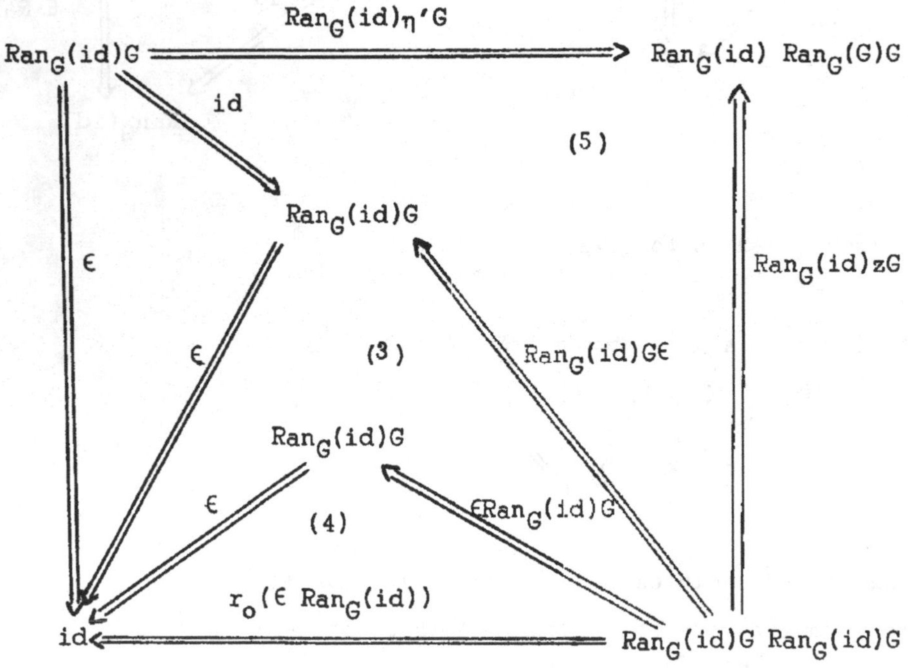

Diagram (5) commutes because it is exactly diagram a) with

$Ran_G(id)$ on the left. Diagram (3) commutes by naturality. Finally, the commutativity of diagram (4) is equivalent by r_o to that of

$$\begin{array}{ccc} Ran_G(id)G\,Ran_G(id) & \xRightarrow{\epsilon\, Ran_G(id)} & Ran_G(id) \\ & \searrow^{\epsilon\, Ran_G(id)} & \Big\Downarrow id \\ & & Ran_G(id) \end{array} ,$$

so diagram (4) commutes. ∎

<u>Remark I.4.1</u>

If a V-functor $\mathbb{B} \xrightarrow{G} \mathbb{A}$ has a V-left adjoint $\mathbb{A} \xrightarrow{F} \mathbb{B}$, then $\mathbb{A} \xrightarrow{Ran_G(id)} \mathbb{B}$ exists and it is preserved by any V-functor $\mathbb{B} \xrightarrow{T} \mathbb{B}'$.

<u>Proof</u>:

It only remains to prove that $Ran_G(id) = F$ is preserved by any functor $\mathbb{B} \xrightarrow{T} \mathbb{B}'$, i.e., that $\mathbb{A} \xrightarrow{TF} \mathbb{B}'$ satisfies the definition of $Ran_G(T)$. Let $FG \xRightarrow{\epsilon} id_{\mathbb{B}}$ and $id_{\mathbb{A}} \xRightarrow{\eta} GF$ be a V-adjunction. Given any V-functor $\mathbb{A} \xrightarrow{H} \mathbb{B}'$ define r_o:

$$r_o \frac{H \Longrightarrow TF}{HG \Longrightarrow T} \qquad \begin{array}{l} r_o(H \xRightarrow{\varphi} TF) = (HG \xRightarrow{\varphi G} TFG \xRightarrow{T\epsilon} T) \\ r_o(HG \xRightarrow{\psi} T) = (H \xRightarrow{H\eta} HGF \xRightarrow{\psi F} TF) , \end{array}$$

then, the result follows in exactly the same way as the first part of Theorem I.4.1. ∎

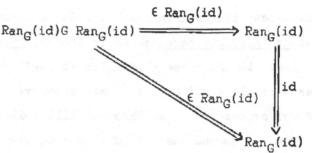

The propositions of our next series are essentially criteria of existence for the right kan extension, which, besides providing a very handy formula to compute it, are the essential deep truths underneath all Adjoint Functor theorems of ordinary set based category theory, where, the end of cotensors in the present formulas (below) can be obtained as a single limit over a comma category.

The first proposition (Kan Theorem) will follow as a corollary from the second one and work done before, but we have decided to give it, first, an independent proof.

Let $\mathbb{C}$,$\mathbb{B}$ be any two V-categories and $\mathbb{C} \xrightarrow{S} \mathbb{B}$ a V-functor. Given a V-functor $\mathbb{C} \xrightarrow{T} \mathbb{A}$ into a cotensored V-category $\mathbb{A}$, consider the V-functor:
$\mathbb{B} \otimes \mathbb{C}^{op} \otimes \mathbb{C} \xrightarrow{\overline{\mathbb{A}}(\mathbb{B}(-,S(-)),T(-))} \mathbb{A}$. Then.

Theorem I.4.2 (Kan Theorem of existence).

If for every $B \in \mathbb{B}$ the end of $\mathbb{C}^{op} \otimes \mathbb{C} \xrightarrow{\overline{\mathbb{A}}(\mathbb{B}(B,S(-)),T(-))} \mathbb{A}$ exists, then, the right kan extension of T along S$\mathbb{B} \xrightarrow{Ran_S(T)} \mathbb{A}$ exists, furthermore, for every $B \in \mathbb{B}$ the formula.

(1) $Ran_S(T)(B) = \int_C \overline{\mathbb{A}}(\mathbb{B}(B,SC),TC)$ holds.

Proof:

Define $Ran_S(T)$ by formula (1), then, by Proposition I.3.2, $Ran_S(T)$ is a V-functor $\mathbb{B} \longrightarrow \mathbb{A}$. Given a V-functor $\mathbb{B} \xrightarrow{F} \mathbb{A}$,

V-natural transformations $FS \xrightarrow{\varphi} T$ and $F \xrightarrow{\psi} \mathrm{Ran}_S(T)$ are the same as V-natural families:

$$I \xrightarrow{\varphi C} \mathbb{A}(FSC, TC) \quad \text{for every } C \in \mathbb{C},$$

$$I \xrightarrow{\psi A} \mathbb{A}(FB, \mathrm{Ran}_S(T)(B)) \text{ for every } B \in \mathbb{B}.$$

To obtain a one to one and onto correspondence between these families we proceed as follows:

Consider the diagram:

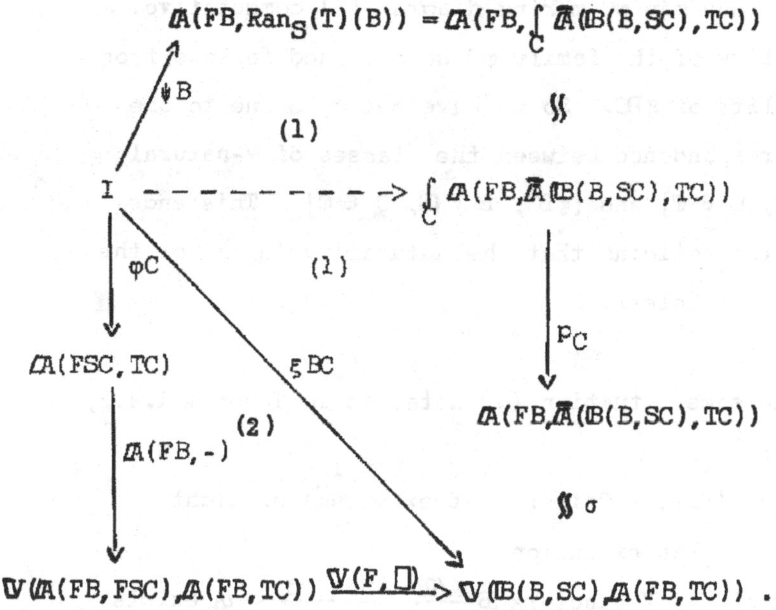

Given a V-natural family $\{\psi B, B \in B\}$, since $\{pC, C \in C\}$ and σ are V-natural, we clearly obtain a V-natural family

$\{\xi BC,\ B \in \mathbb{B},\ C \in \mathbb{C}\}$. Vice versa, if we start with a family ξBC, fixing the B, we can lift it into the end, obtaining a family ψB which is V-natural by Proposition I.3.4. Diagram (1) obviously commutes, and so we have set up a one to one and onto correspondence between the classes of V-natural families $\{\psi B, B \in \mathbb{B}\}$ and $\{\xi BC,\ B \in \mathbb{B},\ C \in \mathbb{C}\}$.

On the other hand, given a V-natural family ξBC, fixing the C, by the representation theorem (see [2]) there is a unique map φC making diagram (2) commutative. The V-naturality of the family φC so obtained follows from the V-naturality of ξBC. So we have set up a one to one and onto correspondence between the classes of V-natural families $\{\varphi C,\ C \in \mathbb{C}\}$ and $\{\xi BC,\ B \in \mathbb{B},\ C \in \mathbb{C}\}$. This ends the proof after noticing that the naturality in F of these correspondences is clear. ∎

With the same situation (or data) as in Theorem I.4.2, we now state:

<u>Theorem I.4.3</u> (Formal Criteria of existence of right kan extension)

The end of the V-functors $\mathbb{B} \xrightarrow{\overline{\mathbb{A}}(\mathbb{B}(-,SC),TC)} \mathbb{A}$ exists if and only if the kan extension of T along S $\mathbb{B} \xrightarrow{\mathrm{Rang}(T)} \mathbb{A}$ exists, and they are equal. In different words, the formula

(1) $\mathrm{Ran}_S(T) = \int_C \overline{\mathbb{A}}(B(-,SC),TC)$

holds and either side exists if and only if the other does.

<u>Proof</u>:

It is clear that the triangle:

$$\begin{array}{ccc} & & \mathbb{A}(FB, \overline{\mathbb{A}}(\mathbb{B}(B,SC),TC)) \\ & \overset{(\theta C)B}{\nearrow} & \\ I & & \wr\!\wr\ \sigma \\ & \underset{\xi BC}{\searrow} & \\ & & \mathbb{V}(\mathbb{B}(B,SC), \mathbb{A}(FB,TC)) \end{array}$$

gives, naturally in F, a one to one and onto correspondence between V-families of V-natural transformations $F \overset{\theta C}{\Longrightarrow} \overline{\mathbb{A}}(\mathbb{B}(-,SC),TC)$ and V-natural families ξBC. Recall, from the proof of Theorem I.4.2 that the later ones are (naturally in F) in one to one and onto correspondence with V-natural transformations $FS \overset{\varphi}{\Longrightarrow} T$. Consider the following two columns of one to one and onto correspondences (naturals in F):

$$\frac{F \overset{\psi}{\Longrightarrow} \int_C \overline{\mathbb{A}}(\mathbb{B}(-,SC),TC)}{\dfrac{F \overset{\theta C}{\Longrightarrow} \overline{\mathbb{A}}(\mathbb{B}(-,SC),TC)}{\dfrac{I \xrightarrow{\xi BC} \mathbb{V}(\mathbb{B}(B,SC),\mathbb{A}(FB,TC)}{FS \overset{\varphi}{\Longrightarrow} T}}} \qquad \frac{F \overset{\psi}{\Longrightarrow} \mathrm{Ran}_S(T)}{\dfrac{FS \overset{\varphi}{\Longrightarrow} T}{\dfrac{I \xrightarrow{\xi BC} \mathbb{V}(\mathbb{B}(B,SC),\mathbb{A}(FB,TC))}{F \overset{\theta C}{\Longrightarrow} \overline{\mathbb{A}}(\mathbb{B}(B,SC),TC)}}}$$

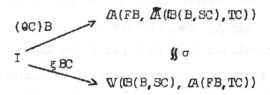

The left column proves that the end satisfies the definition of the right kan extension, while the right column proves that the right kan extension satisfies the definition of the end. ∎

When the hypotheses of Theorem I.4.2 are satisfied we say that $Ran_S(T)$ is a <u>pointwise Kan extension</u>. From formula (1) in Theorem I.4.2 it follows then that pointwise Kan extensions are exactly those for which the end of V-functors in formula (1) of Theorem I.4.3 is a pointwise end. It follows then from the Lemma in page 14 that right kan extensions of V-functors with codomain $\mathbb{V}$ are always pointwise, from which it is not difficult to prove that extensions preserved by all the representables are necessarily pointwise. This result follows anyhow from Proposition I.3.3.

<u>Proposition I.4.3</u>

Let
$$\begin{array}{ccc} \mathbb{C} & \xrightarrow{S} & \mathbb{B} \\ & \searrow^{T} & \\ & & \mathbb{A} \end{array}$$
any three V-categories and V-functors, where $\mathbb{A}$ is cotensored. Suppose $\mathbb{B} \xrightarrow{Ran_S(T)} \mathbb{A}$ exists. Then, $Ran_S(T)$ is pointwise if and only if it is preserved by all the representables. ∎

If $\mathbb{A}$ is tensored as well as cotensored, from the above and Proposition I.4.2 it follows that any extension with codomain $\mathbb{A}$ is necessarily pointwise. Explicitly, we have:

<u>Proposition I.4.4</u>

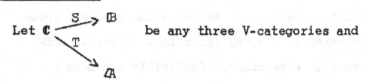

Let $\mathbb{C} \xrightarrow{S} \mathbb{B}$, $\mathbb{C} \xrightarrow{T} \mathbb{A}$ be any three V-categories and V-functors where $\mathbb{A}$ is tensored and cotensored, then $\mathbb{B} \xrightarrow{Ran_S(T)} \mathbb{A}$ exists if and only if for every $B \in \mathbb{B}$ the end of $\mathbb{C}^{op} \otimes \mathbb{C} \xrightarrow{\overline{\mathbb{A}}(\mathbb{B}(B,S-),T(-))} \mathbb{A}$ exists, and the formula

$$Ran_S(T)(B) = \int_C \overline{\mathbb{A}}(\mathbb{B}(B,SC),TC) \text{ holds.}$$ ∎

Note that the tensority assumption can be weakened into preservation of ends of V-functors by the representables.

Finally, it follows from Remark I.4.1 and Proposition I.4.3 that the right kan extension in Theorem I.4.1 is always pointwise (when G has a V-left adjoint). The formal criteria of existence of left adjoint can be expressed, hence, in the following form:

<u>Theorem I.4.4</u> (Benabou [5])

Given any V-functor $\mathbb{B} \xrightarrow{G} \mathbb{A}$, where $\mathbb{B}$ is cotensored, G has a V-left adjoint $\mathbb{A} \xrightarrow{F} \mathbb{B}$ if and only if for every

$A \in \mathbb{A}$, $\int_B \overline{\mathbb{B}}(\mathbb{A}(A,GB),B)$ exist (in $\mathbb{B}$) and is preserved by G. Moreover, the formula

$$FA = \int_B \overline{\mathbb{B}}(\mathbb{A}(A,GB),B) \quad \text{holds.}$$ ∎

We finish this section by establishing the well known fact that kan extensions along inclusions of V-full subcategories are real extensions. Explicitly and more generally:

Proposition I.4.5

Given any V-full-and-faithful V-functor $\mathbb{C} \xrightarrow{S} \mathbb{B}$, the right kan extension $\mathbb{B} \xrightarrow{Ran_S(T)} \mathbb{A}$ ($\mathbb{A}$ cotensored) of any V-functor $\mathbb{C} \xrightarrow{T} \mathbb{A}$ is such that the V-natural transformation $Ran_S(T)S \overset{\epsilon}{\Longrightarrow} T$ is an isomorphism.

Proof:

We have

$$\begin{array}{ccc} Ran_S(T)S & = & \int_C \overline{\mathbb{A}}(\mathbb{B}(S(-),SC),TC) \\ \Big\Downarrow \epsilon & & \Big\Downarrow \int_C \overline{\mathbb{A}}(S,\square) \\ T = Ran_{id}(T) & = & \int_C \overline{\mathbb{A}}(\mathbb{C}(-,C),TC) \end{array},$$

the two equalities by Theorem I.4.3 . ∎

Remark I.4.2

For a general $\mathbb{C} \xrightarrow{S} \mathbb{B}$ (not necessarily V-full-and-faithful) if $Ran_S(T)$ is pointwise, then $Ran_S(T)SC \xrightarrow{\epsilon C} TC$ is an isomorphism for any $C \in \mathbb{C}$ such that $\mathbb{C}(C,-) \overset{S}{\Longrightarrow} \mathbb{B}(SC,S(-))$ is an isomorphism. ∎

Section 5 The V-Yoneda lemma

Given any two V-categories $\mathbb{C}$, $\mathbb{A}$ and a V-functor $\mathbb{C} \xrightarrow{T} \mathbb{A}$, it is clear that the right kan extension of T along the identity, $\mathbb{C} \xrightarrow{id} \mathbb{C}$, exists and is equal to T: $Ran_{id}(T) = T$. As we have already seen in Proposition I.4.5, in view of Theorem I.4.3, this obvious observation, (under different interpretations) means a variety of facts (or results) in category theory. In particular, it means exactly the V-Yoneda Lemma.

Proposition I.5.1

For any V-functor $\mathbb{C} \xrightarrow{T} \mathbb{A}$, $\mathbb{A}$ cotensored, the end of V-functors in the following formula exists and the formula holds:

$$Ran_{id}(T) = T = \int_C \mathbb{A}\,(\mathbb{C}(-,C),TC) .$$

Dually, if $\mathbb{A}$ is tensored (By Theorem I.4.3 dual),

$$Lan_{id}(T) = T = \int^C \mathbb{C}(C,-) \otimes_{\mathbb{A}} TC . \qquad ∎$$

Given any two V-functors between any two V-categories, $\mathbb{C} \xrightarrow{T} \mathbb{A}$, $\mathbb{C} \xrightarrow{H} \mathbb{A}$, consider the V-bifunctor $\mathbb{C}^{op} \otimes \mathbb{C} \xrightarrow{\mathbb{A}(T(-),H(-))} \mathbb{V}$. Assume its end exists: $\int_C \mathbb{A}(TC,HC) \in \mathbb{V}$. We have then, by definition, a one to one and onto correspondence between morphisms

$I \longrightarrow \int_C \mathbb{A}(TC,HC)$ and V-natural families $I \xrightarrow{\varphi C} \mathbb{A}(TC,HC)$. But these are the same as V-natural transformations $T \xRightarrow{\varphi} H$. So, the underlying set of $\int_C \mathbb{A}(TC,HC)$, $\mathbb{V}_o(I, \int_C \mathbb{A}(TC,HC))$, has as its elements exactly all the V-natural transformations between T and H. For this reason, and using the notation $\mathbb{A}^{\mathbb{C}}$ in a purely symbolic way, we write:

$$(1) \qquad \mathbb{A}^{\mathbb{C}}(T,H) = \int_C \mathbb{A}(TC,HC) \quad .$$

We see then, that the existence of this end means that the class of V-natural transformations between T and H is actually a set, and that it can be lifted into $\mathbb{V}$.

Consider now any V-functor $\mathbb{C} \xrightarrow{T} \mathbb{V}$. Since ends of V-functors with codomain $\mathbb{V}$ are always pointwise, the end in Proposition I.5.1 is in this case pointwise, and so, for any $D \in \mathbb{C}$ we have:

$$TD = \int_C \mathbb{V}(\mathbb{C}(D,C),TC) = \mathbb{V}^{\mathbb{C}}(\mathbb{C}(D,-),T) \ .$$

That is, in this case, the end (1) above always exists and is TD. The projections are $p_C = G_o(T_{D,C})$.

<u>Proposition I.5.2</u> (V-Yoneda Lemma).

Given any V-functor $\mathbb{C} \xrightarrow{T} \mathbb{V}$, for any object $D \in \mathbb{C}$, the class of V-natural transformations between $\mathbb{C}(D,-)$ and T

is the underlying set of TD. ∎

In particular, when $T = \mathbb{C}(D',-)$, we have:

$$\mathbb{C}(D',D) = \int_C \mathbb{V}(\mathbb{C}(DC),\mathbb{C}(D'D)) = \mathbb{V}^{\mathbb{C}}(\mathbb{C}(D,-),\mathbb{C}(D'-)) \quad .$$

The formulas in Proposition I.5.1 are an expression of any V-functor $\mathbb{C} \xrightarrow{T} \mathbb{A}$ as an end of cotensors of contravariant representable functors and as a coend of tensors of covariant representable functors (called generalized representables in [10]). This formula means also a result for which we still have no name in this stage of the paper. However, it seems convenient to do some formal manipulation now. If $\mathbb{V} = \mathbb{A}$, both formulas apply, and recalling the definition of pointwise cotensors of V-functors (in this case tensors) we obtain: (reading from bottom to top)

$$(1) \qquad T = \int^C \mathbb{V}^{\mathbb{C}}(\mathbb{C}(C,-),T) \otimes_{\mathbb{V}^{\mathbb{C}}} \mathbb{C}(C,-)$$

$$= \int^C TC \otimes_{\mathbb{V}^{\mathbb{C}}} \mathbb{C}(C,-) = \int^C TC \otimes \mathbb{C}(C,-) =$$

$$= \int^C \mathbb{C}(C,-) \otimes TC = T \quad .$$

CHAPTER II

V-MONADS

Section 1. Semantics-Structure (meta) Adjointness

Given a V-category $\mathbb{A}$, recall that a V-monad in $\mathbb{A}$ is a V-endofunctor $\mathbb{A} \xrightarrow{T} \mathbb{A}$ together with a pair of V-natural transformations $TT \xRightarrow{\mu} T$ and $id_{\mathbb{A}} \xRightarrow{\eta} T$, μ is associative and η is a left and right unit for μ in the sense that the following diagrams commute:

$$\begin{array}{ccc} TTT & \xRightarrow{\mu T} & TT \\ {\scriptstyle T\mu}\Downarrow & & \Downarrow{\scriptstyle \mu} \\ TT & \xRightarrow{\mu} & T \end{array} \quad , \quad \begin{array}{ccc} T & \xRightarrow{\eta T} & TT \\ & {\scriptstyle id}\searrow & \Downarrow{\scriptstyle \mu} \\ & & T \end{array} \quad \text{and} \quad \begin{array}{ccc} T & \xRightarrow{T\mu} & TT \\ & {\scriptstyle id}\searrow & \Downarrow{\scriptstyle \mu} \\ & & T \end{array}$$

We write $\mathbb{T} = (T, \mu, \eta)$ and call μ the multiplication, and η the unit. A morphism of monads $\mathbb{T} \xrightarrow{\varphi} \mathbb{T}'$ is a V-natural transformation $T \xRightarrow{\varphi} T'$ such that the diagram

$$\begin{array}{ccc} TT & \xRightarrow{\varphi\varphi} & T'T' \\ {\scriptstyle \mu}\Downarrow & & \Downarrow{\scriptstyle \mu'} \\ T & \xRightarrow{\varphi} & T' \end{array} \quad \text{and} \quad \begin{array}{ccc} T & \xRightarrow{\varphi} & T' \\ {\scriptstyle \mu}\searrow & & \swarrow{\scriptstyle \mu'} \\ & id_{\mathbb{A}} & \end{array} \quad \text{commute} .$$

V-monads in $\mathbb{A}$ with morphisms of monads between them form a (meta) category that we denote $\mathcal{M}(\mathbb{A})$.

• The V-category of algebras

A **$\mathbb{T}$-algebra** is an object $A \in \mathbb{A}$ together with a $\mathbb{T}$-algebra structure, that is, a morphism $TA \xrightarrow{a} A$, associative and for which ηA is a unit, in the sense that the diagrams

$$\begin{array}{ccc} TTA & \xrightarrow{Ta} & TA \\ \downarrow{\scriptstyle \mu A} & & \downarrow{\scriptstyle a} \\ TA & \xrightarrow{a} & A \end{array} \quad \text{and} \quad \begin{array}{ccc} A & \xrightarrow{\eta A} & TA \\ & {\scriptstyle id}\searrow & \swarrow{\scriptstyle a} \\ & & A \end{array} \quad \text{commute.}$$

We write $\bar{A} = (A, a)$ and call A the underlying object.

A *morphism of algebras* $\bar{A} \xrightarrow{f} \bar{B}$ is a map $A \xrightarrow{f} B$ in $\mathbb{A}$ such that the diagram

$$\begin{array}{ccc} TA & \xrightarrow{Tf} & TB \\ \downarrow{\scriptstyle a} & & \downarrow{\scriptstyle b} \\ A & \xrightarrow{f} & B \end{array}$$

commutes.

$\mathbb{T}$-algebras and morphisms of algebras form a category $\mathbb{A}^{\mathbb{T}}$ provided with a functor $\mathbb{A}^{\mathbb{T}} \xrightarrow{U^{\mathbb{T}}} \mathbb{A}$, $U^{\mathbb{T}}\bar{A} = A$, $U^{\mathbb{T}}f = f$. Assume now that $\mathbb{V}$ has equalizers, then $\mathbb{A}^{\mathbb{T}}$ is a V-category and $U^{\mathbb{T}}$ a V-functor by defining $\mathbb{A}^{\mathbb{T}}(\bar{A},\bar{B}) \xrightarrow{U^{\mathbb{T}}} \mathbb{A}(A,B)$ to be a V-equalizer of the pair of maps:

$$\begin{array}{ccc} \mathbb{A}(A,B) & \xrightarrow{\mathbb{A}(a,\square)} & \mathbb{A}(TA,B) \\ & {\scriptstyle T}\searrow \quad \nearrow{\scriptstyle \mathbb{A}(\square,b)} & \\ & \mathbb{A}(TA,TB) & \end{array}$$

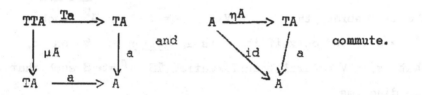

$U^{\mathbb{T}}$ is obviously V-faithful and we call it the forgetful functor.

• Proposition II.1.1

Given a V-functor $\mathbb{C} \xrightarrow{S} \mathbb{A}$; S admits a lifting into the $\mathbb{T}$-algebras, that is, a V-functor $\mathbb{C} \xrightarrow{\bar{S}} \mathbb{A}^{\mathbb{T}}$ such that $U^{\mathbb{T}}\bar{S} = S$ if and only if there is an action of $\mathbb{T}$ on S, that is, a V-natural transformation $TS \xRightarrow{s} S$ such that the diagrams

$$\begin{array}{ccc} TTS & \xRightarrow{Ts} & TS \\ \Big\Downarrow{\scriptstyle \mu S} & & \Big\Downarrow{\scriptstyle s} \\ TS & \xRightarrow{s} & S \end{array} \qquad \text{and} \qquad \begin{array}{ccc} S & \xRightarrow{\eta S} & TS \\ & {\scriptstyle id}\searrow & \Big\Downarrow{\scriptstyle s} \\ & & S \end{array} \qquad \text{commute.}$$

Furthermore: the correspondence between $\mathbb{T}$-actions and liftings is one to one and onto.

Proof:

Suppose there is an action $TS \xrightarrow{s} S$ of $\mathbb{T}$ on S. Define $\mathbb{C} \xrightarrow{\bar{S}} \mathbb{A}^{\mathbb{T}}$ by $\bar{S}(C) = (TSC \xrightarrow{sC} SC)$. That $\bar{S}(C)$ is a $\mathbb{T}$-algebra follows trivially from the fact that s is a $\mathbb{T}$-action. The V-structure of $\bar{S}$ is gotten in the following way:

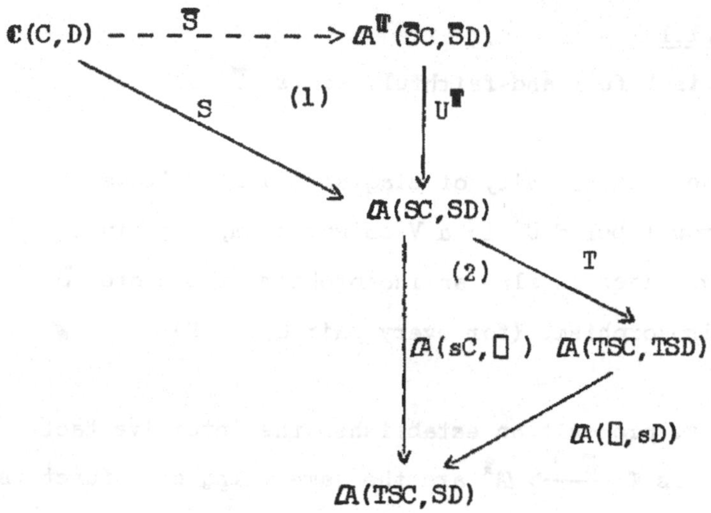

The V-naturality of s says that S equalizes the two maps of diagram (2), so there is a (unique) map $\bar{S}$ making diagram (1) commutative. It can be seen that $\bar{S}$ is actually a V-functor; and diagram (1) commutative means exactly that it is a lifting of S.

Conversely, suppose there is a lifting $\mathbb{C} \xrightarrow{\bar{S}} \mathbb{A}^{\mathbb{T}}$, $\bar{S}(C) = (TSC \xrightarrow{sC} SC)$. Define $TS \xRightarrow{s} S$ by the equation $sC = sC$. Then, that s is a $\mathbb{T}$-action follows trivially from the fact that $\bar{S}C$ is a $\mathbb{T}$-algebra for every $C \in \mathbb{C}$. Diagram (1) now commutes because $\bar{S}$ is a lifting, so S equalizes the two maps of diagram (2), that is, s is V-natural. Finally, it is clear that the correspondences are inverses each of the other. ∎

• Remark II.1.1

If S is V-full-and-faithful, so is $\bar{S}$.

Proof:

From the commutativity of diagram (1) it follows that the arrow labeled $U^{\mathbb{T}}$ is a V-epimorphism, so, since it is a V-equalizer is also an isomorphism, therefore $\bar{S}$ is also an isomorphism. (for every pair $C, D \in \mathbb{C}$) ∎

The above proposition establishes the intuitive fact that V-functors $\mathbb{C} \xrightarrow{\bar{S}} \mathbb{A}^{\mathbb{T}}$ are the same thing as V-functors $\mathbb{C} \xrightarrow{S} \mathbb{A}$ together with a V-natural $\mathbb{T}$-algebra structure $TS \overset{s}{\Longrightarrow} S$, that is, together with an action. A V-natural transformation from $\bar{S}$ into any other $\mathbb{C} \xrightarrow{\bar{H}} \mathbb{A}^{\mathbb{T}}$ should then be just a V-natural transformation $S \overset{\varphi}{\Longrightarrow} H$ such that the diagram

$$\begin{array}{ccc} TS & \overset{T\varphi}{\Longrightarrow} & TH \\ {\scriptstyle s}\Downarrow & & \Downarrow{\scriptstyle h} \\ S & \overset{\varphi}{\Longrightarrow} & H \end{array} \qquad (1)$$

commutes. This is actually true and it follows easily from the fact that $U^{\mathbb{T}}$, being V-faithful, reflects V-naturality.

The identity functor $\mathbb{A}^{\mathbb{T}} \xrightarrow{id} \mathbb{A}^{\mathbb{T}}$ is the lifting of $U^{\mathbb{T}}$, and so there is an action $TU^{\mathbb{T}} \overset{u}{\Longrightarrow} U^{\mathbb{T}}$, $u\bar{A} = a$.

Also, since $TT \overset{\mu}{\Longrightarrow} T$ is an action of $\mathbb{T}$ on T, there is a lifting of T into the $\mathbb{T}$-algebras $\mathbb{A} \xrightarrow{F^{\mathbb{T}}} \mathbb{A}^{\mathbb{T}}$, $U^{\mathbb{T}}F^{\mathbb{T}} = T$, $F^{\mathbb{T}}A = (TTA \xrightarrow{\mu A} A)$. It is clear that $uF^{\mathbb{T}} = \mu$. One of the equations in the definition of an action is exactly diagram (1) above for u, and so there is a V-natural transformation $F^{\mathbb{T}}U^{\mathbb{T}} \overset{\epsilon}{\Longrightarrow} id$, $U^{\mathbb{T}}\epsilon = u$, that, together with $id \overset{\eta}{\Longrightarrow} U^{\mathbb{T}}F^{\mathbb{T}}$, establishes the fact that $F^{\mathbb{T}}$ is V-left adjoint to $U^{\mathbb{T}}$. The triangular equation

$$U^{\mathbb{T}} \overset{\eta U^{\mathbb{T}}}{\Longrightarrow} U^{\mathbb{T}}F^{\mathbb{T}}U^{\mathbb{T}} \overset{U^{\mathbb{T}}\epsilon}{\Longrightarrow} U^{\mathbb{T}} = U^{\mathbb{T}} \overset{id}{\Longrightarrow} U^{\mathbb{T}}$$

is the other equation in the definition of action, and

$$F^{\mathbb{T}} \overset{F^{\mathbb{T}}\eta}{\Longrightarrow} F^{\mathbb{T}}U^{\mathbb{T}}F^{\mathbb{T}} \overset{\epsilon F^{\mathbb{T}}}{\Longrightarrow} F^{\mathbb{T}} = F^{\mathbb{T}} \overset{id}{\Longrightarrow} F^{\mathbb{T}}$$

taken downstairs is

$$T \overset{T\eta}{\Longrightarrow} TT \overset{\mu}{\Longrightarrow} T = T \overset{id}{\Longrightarrow} T.$$

So, we have just proven the following:

• <u>Proposition II.1.2</u>

$F^{\mathbb{T}}$ is V-left adjoint to $U^{\mathbb{T}}$ and the V-monad $(U^{\mathbb{T}}F^{\mathbb{T}}, U^{\mathbb{T}} \epsilon F^{\mathbb{T}}, \eta)$ is equal to $\mathbb{T}$. (recall that given any pair of V-adjoints $(\epsilon,\eta): F \dashv_V G$, the triple $(GF, G\epsilon F, \eta)$ is a V-monad) ∎

We call the V-functor $F^{\mathbb{T}}$ the <u>free functor</u> and a $\mathbb{T}$-algebra of the form $F^{\mathbb{T}}A$ a <u>free algebra</u>.

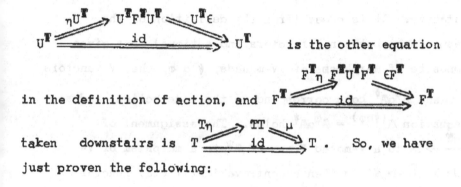

Given a morphism of monads $\mathbb{T}' \xrightarrow{\varphi} \mathbb{T}$ it is trivial to see that $T'U^{\mathbb{T}} \xRightarrow{\varphi U^{\mathbb{T}}} TU^{\mathbb{T}} \xRightarrow{u} U^{\mathbb{T}}$ is an action of $\mathbb{T}'$ on $U^{\mathbb{T}}$, and so, there is a V-functor, denoted $\mathbb{A}^{\varphi}$, which makes the triangle

$$\begin{array}{ccc} \mathbb{A}^{\mathbb{T}} & \xrightarrow{\mathbb{A}^{\varphi}} & \mathbb{A}^{\mathbb{T}'} \\ & {\scriptstyle U^{\mathbb{T}}} \searrow \quad \swarrow {\scriptstyle U^{\mathbb{T}'}} & \\ & \mathbb{A} & \end{array}$$

commutative. It is clear (from the definition of the correspondence between V-functors and actions) that given a composite of morphisms of V-monads, $\psi \circ \varphi$, the V-functors $\mathbb{A}^{(\psi\circ\varphi)}$ and $\mathbb{A}^{\varphi}\circ\mathbb{A}^{\psi}$ both correspond to the same action, and so, the equation $\mathbb{A}^{(\psi\circ\varphi)} = \mathbb{A}^{\varphi}\circ\mathbb{A}^{\psi}$ holds. The assignment of $\mathbb{A}^{T} \xrightarrow{U^{\mathbb{T}}} \mathbb{A}$ to a V-monad $\mathbb{T}$ and of $\mathbb{A}^{\varphi}$ to a morphism of V-monads $\mathbb{T} \xrightarrow{\varphi} \mathbb{T}'$ is then a contravariant (meta) functor between $\mathcal{M}o(A)$ and the (meta) comma category $(\mathbb{C}at,\mathbb{A})$.

$$\mathcal{M}o(A)^{op} \xrightarrow{\mathcal{G}} (\mathbb{C}at,\mathbb{A})$$

In this notation, we can write the one to one and onto correspondence of Proposition II.1.1 by:

$$\frac{S \longrightarrow \mathcal{G}(\mathbb{T})}{TS \Longrightarrow S}$$

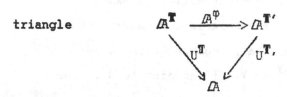

Where the above arrow is understood to be a map in $(\mathbb{C}at, \mathbb{A})$ and the above double arrow an action of $\mathbb{T}$ on S.

If $\mathbb{T} \xrightarrow{\varphi} \mathbb{T}'$ is a morphism of V-monads and $T'S \xRightarrow{s} S$ is an action of $\mathbb{T}'$ on S; the composite $TS \xRightarrow{\varphi S} T'S \xRightarrow{s} S$ is an action of $\mathbb{T}$ on S, and it is not difficult to check the following fact:

• Proposition II.1.3

The one to one and onto correspondence

$$\frac{S \longrightarrow \mathfrak{S}(\mathbb{T})}{TS \Longrightarrow S}$$

is natural in T with respect to morphisms of V-monads. ∎

The Codensity V-monad

Given a V-functor $\mathbb{C} \xrightarrow{S} \mathbb{A}$, the right Kan extension of S along itself, $\mathbb{A} \xrightarrow{\mathrm{Ran}_S(S)} \mathbb{A}$, has a structure of V-monad given by:

$$(1) \qquad r_o \; \frac{S \xRightarrow{id} S}{id \xRightarrow{\eta} \mathrm{Ran}_S(S)}$$

and

$$r_o \frac{\mathrm{Ran}_S(S) \xrightarrow{\mathrm{id}} \mathrm{Ran}_S(S)}{}$$

$$(2) \quad r_o \frac{\mathrm{Ran}_S(S)\mathrm{Ran}_S(S)S \xrightarrow{\mathrm{Ran}_S(S)\epsilon} \mathrm{Ran}_S(S)S \xrightarrow{\epsilon} S}{\mathrm{Ran}_S(S)\mathrm{Ran}_S(S) \xrightarrow{\mu} \mathrm{Ran}_S(S)}$$

We write $\mathbb{T}_S = (\mathrm{Ran}_S(S), \mu, \eta)$ and call it the <u>codensity V-monad</u>. If it exists, we say that S <u>admits</u> a codensity V-monad. We say that S is <u>tractable</u> if, furthermore, $\mathrm{Ran}_S(S)$ is preserved by the representables. It follows then (Proposition I.4.3) that for tractable S and cotensored $\mathbb{A}$, $\mathrm{Ran}_S(S)$ is always pointwise.

That $\mathbb{T}_S$ is actually a V-monad can be seen as follows:
Associativity:

The commutativity of the diagram

$$\begin{array}{ccc} \mathrm{Ran}_S(S)\mathrm{Ran}_S(S)\mathrm{Ran}_S(S) & \xrightarrow{\mu\mathrm{Ran}_S(S)} & \mathrm{Ran}_S(S)\mathrm{Ran}_S(S) \\ \Big\Downarrow \mathrm{Ran}_S(S)\mu & & \Big\Downarrow \mu \\ \mathrm{Ran}_S(S)\mathrm{Ran}_S(S) & \xrightarrow{\mu} & \mathrm{Ran}_S(S) \end{array}$$

is equivalent by r_o to that of the exterior of the diagram:

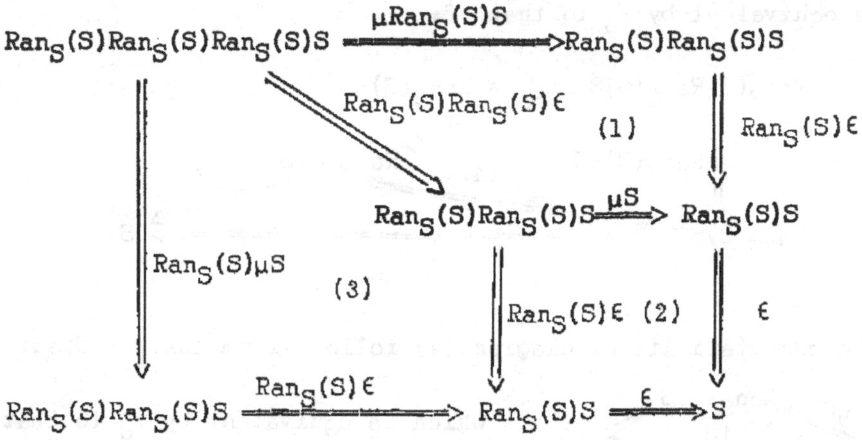

The commutativity of diagram (2) is equivalent by r_o to that of:

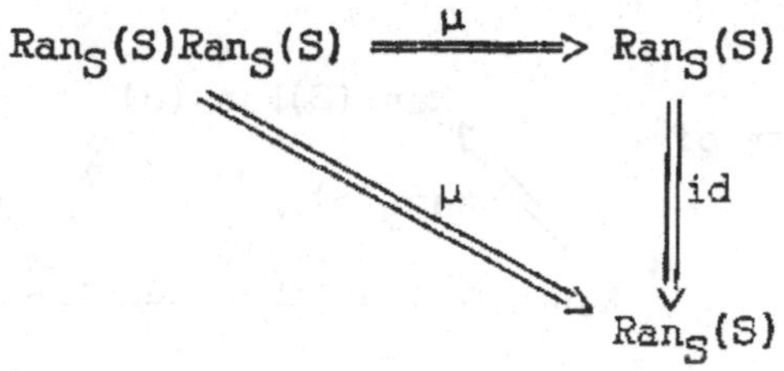

So diagram (2) commutes. Diagram (3) commutes because it is exactly diagram (2) with $Ran_S(S)$ on the left. Finally, diagram (1) commutes by naturality.

Right unit.

The commutativity of

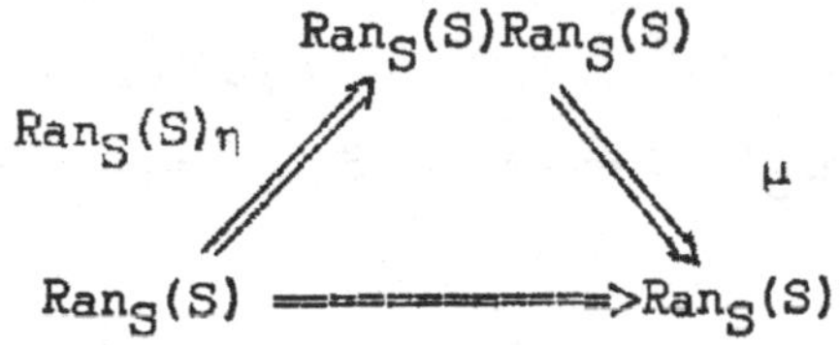

is equivalent by r_o to that of:

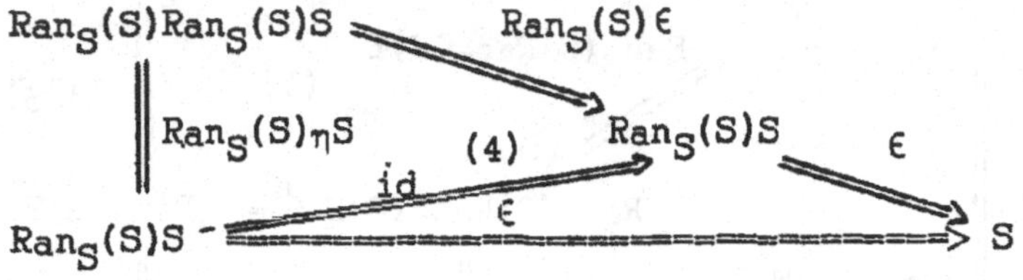

The commutativity of diagram (4) follows from that of diagram

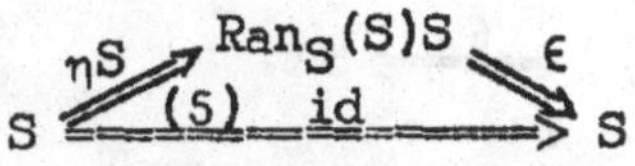

which is equivalent by r_o to that of

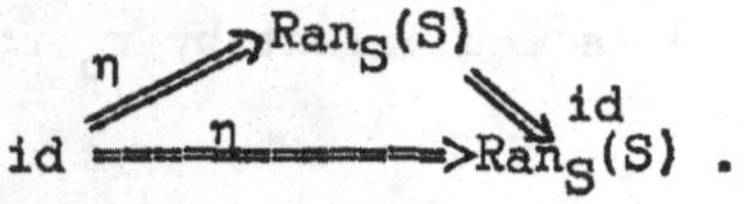

Left unit.

The commutativity of

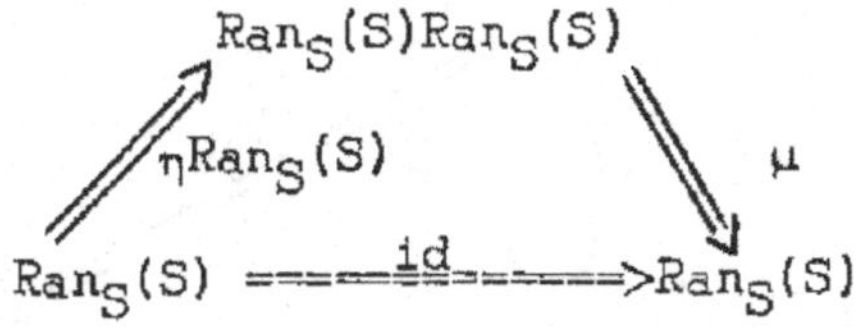

is equivalent by r_o to that of the exterior of diagram

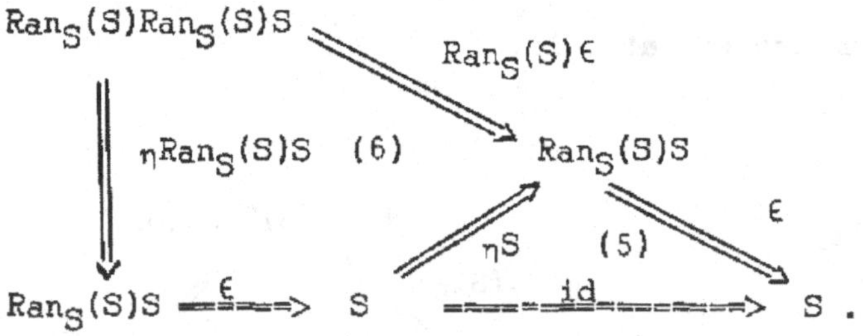

We know already that diagram (5) commutes, and diagram (6) commutes by naturality.

Proposition II.1.4

Given any other V-monad $\mathbf{T}$ in A, $\mathbf{T} = (T,\mu',\eta')$, actions of $\mathbf{T}$ on S and morphism of V-monads $\mathbf{T} \longrightarrow \mathbf{T}_S$ correspond to each other under r_o :

$$TS \Longrightarrow S$$

$$r_o \overline{\phantom{\hspace{4cm}}}$$

$$T \Longrightarrow Ran_S(S) \quad .$$

Proof:

Let $T \stackrel{\varphi}{\Longrightarrow} Ran_S(S)$ be any V-natural transformation the diagrams expressing the fact that $r_o(\varphi)$ is an action are:

a) and the exterior of diagram:

b)

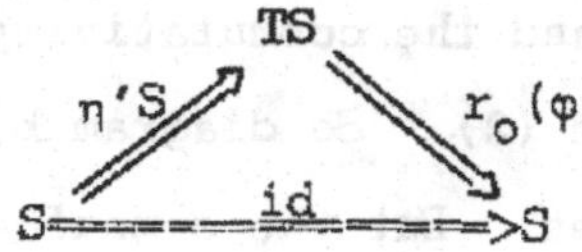

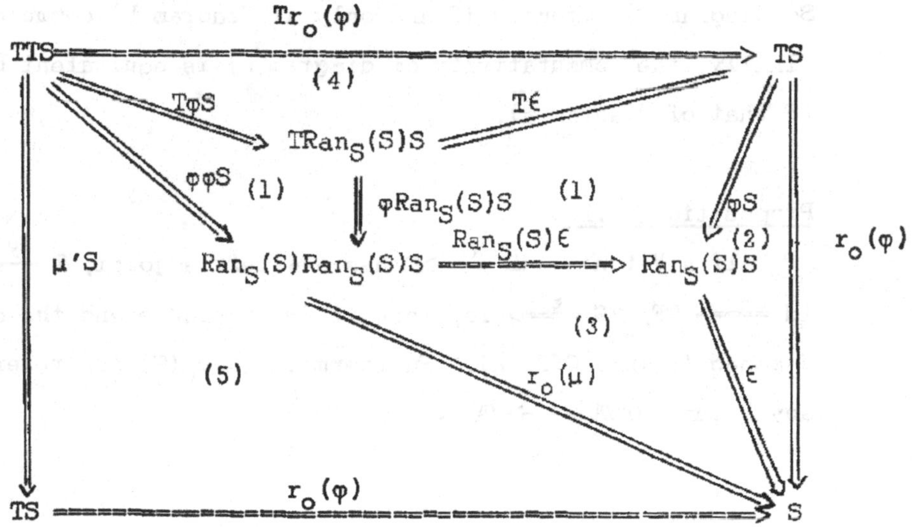

The diagrams expressing the fact that φ is a morphism of V-monads are:

$$\text{c)}\quad \begin{array}{ccc} & T & \\ {\scriptstyle\eta'}\nearrow & & \searrow{\scriptstyle\varphi} \\ \mathrm{id} & \xrightarrow{\ \eta\ } & \mathrm{Ran}_S(S) \end{array} \qquad \text{and} \quad \text{d)}\quad \begin{array}{ccc} TT & \xrightarrow{\ \varphi\varphi\ } & \mathrm{Ran}_S(S)\mathrm{Ran}_S(S) \\ \Big\Downarrow{\scriptstyle\mu'} & & \Big\Downarrow{\scriptstyle\mu} \\ T & \xrightarrow{\ \varphi\ } & \mathrm{Ran}_S(S) \end{array}$$

In diagram b); diagrams (1) commute by naturality, the commutativity of diagram (2) is equivalent by r_o to that of

$$\begin{array}{ccc} T & \xrightarrow{\ \varphi\ } & \mathrm{Ran}_S(S) \\ & {\scriptstyle\varphi}\searrow & \Big\Downarrow{\scriptstyle\mathrm{id}} \\ & & \mathrm{Ran}_S(S) \end{array}\ ,$$

so diagram (2) commutes.

Diagram (3) commutes by definition of μ and the commutativity of diagram (4) follows from that of diagram (2). So diagram b) commutes if and only if diagram (5) commutes. But the commutativity of diagram (5) is equivalent by r_o to that of diagram d). So diagram d) commutes if and only if diagram b) commutes. Finally, the commutativity of diagram c) is equivalent by r_o to that of diagram a). ∎

Proposition II.1.5

If a V-functor $\mathbb{B} \xrightarrow{G} \mathbb{A}$ has a V-left adjoint, $\mathbb{A} \xrightarrow{F} \mathbb{B}$, $\mathrm{id} \xRightarrow{\eta} GF$, $FG \xRightarrow{\epsilon} \mathrm{id}$, then it is tractable and the codensity V-monad is $(GF, G\epsilon F, \eta)$. Furthermore, $\mathrm{Ran}_G(G)$ is preserved by any V-functor $\mathbb{A} \xrightarrow{T} \mathbb{A}'$.

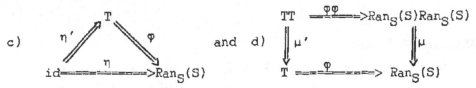

<u>Proof</u>:

By Theorem I.4.1 we can assume that $Ran_G(G)$ is GF, with an r_o given by the formulas:

$$r_o(H \xRightarrow{\varphi} GF) = (HG \xRightarrow{\varphi G} GFG \xRightarrow{G\epsilon} G)$$

$$r_o(HG \xRightarrow{\psi} G) = (H \xRightarrow{H\eta} HGF \xRightarrow{\psi F} GF)$$

Then; it is trivial to see that definitions (1) and (2)(pages 68 and 67) produce the V-monad $(GF, G\epsilon F, \eta)$. Using Remark I.4.1 the following two equalities finish the proof:

$T\ Ran_G(G) = TG\ Ran_G(id) = Ran_G(TG)$ ∎

• <u>Theorem II.1.1</u>

Given a V-functor $\mathbb{C} \xrightarrow{S} \mathbb{A}$ which admits a codensity monad, for every V-monad $\mathbb{T} \in \mathcal{M}(\mathbb{A})$, there is a natural in $\mathbb{T}$ one to one and onto correspondence between morphisms of V-monads $\mathbb{T} \longrightarrow \mathbb{T}_S$ and V-functors $\mathbb{C} \longrightarrow \mathbb{A}^{\mathbb{T}}$ making the triangle

$$\begin{array}{ccc} \mathbb{C} & \longrightarrow & \mathbb{A}^{\mathbb{T}} \\ & {\scriptstyle S}\searrow \quad \swarrow{\scriptstyle U^{\mathbb{T}}} & \\ & \mathbb{A} & \end{array}$$

commutative. That is, maps $S \longrightarrow \mathcal{G}(\mathbb{T})$ in $(\mathbb{C}at, \mathbb{A})$.

As usual, we indicate this by $\dfrac{S \longrightarrow \mathcal{G}(T)}{\mathbb{T} \longrightarrow \mathbb{T}_S}$

<u>Proof</u>:

Immediate from Propositions II.1.3 and II.1.4 ∎

Let $\mathcal{Z}_r(\mathbb{C}at, \mathbb{A})$ the full (meta) sub-category of $(\mathbb{C}at, \mathbb{A})$ whose objects are the V-functors admitting a codensity V-monad. From Propositions II.1.2 and II.1.5 we know that the semantics (meta) functor $\mathfrak{S}$ takes its values in $\mathcal{Z}_r(\mathbb{C}at, \mathbb{A})$. The assignment of $\mathbb{T}_S \in \mathcal{M}(\mathbb{A})$ to a V-functor $\mathbb{C} \xrightarrow{S} \mathbb{A}$ becomes then, by Theorem II.1.1, a contravariant (meta) functor, denoted $\check{\mathfrak{S}}$, in such a way that the one to one and onto correspondence (in Theorem II.1.1) is also natural in S. $\check{\mathfrak{S}}$ is then a left adjoint to semantics, and it is called <u>structure</u>.

$$\mathcal{M}(\mathbb{A})^{op} \underset{\check{\mathfrak{S}}}{\overset{\mathfrak{S}}{\rightleftarrows}} \mathcal{Z}_r(\mathbb{C}at, \mathbb{A})$$

Given a V-functor $\mathbb{C} \xrightarrow{S} \mathbb{A}$ in $\mathcal{Z}_r(\mathbb{C}at, \mathbb{A})$, the codensity V-monad $\mathbb{T}_S = \check{\mathfrak{S}}(S)$ is the structure of S.

Notice again that the correspondence in Proposition II.1.1 is essentially an identity. A V-functor $\mathbb{C} \xrightarrow{\bar{S}} \mathbb{A}^{\mathbb{T}}$ is a function on the objects, $\bar{S}C \in \mathbb{A}^{\mathbb{T}}$, $\bar{S}C = (TSC \xrightarrow{sC} SC)$, plus a V-structure. A $\mathbb{T}$-action on S is a family of maps $TSC \xrightarrow{sC} SC$ such that sC is a $\mathbb{T}$-algebra structure on SC, plus the V-naturality requirement. We see clearly then that in both cases we have the same data, i.e., a family of arrows sC, $C \in \mathbb{C}$, the V-functor structure in the first case being equivalent to the V-naturality in the second.

The (meta) adjunction:
$$\frac{S \longrightarrow \mathcal{G}(\mathbb{T})}{\mathbb{T} \longrightarrow \tilde{\mathcal{G}}(S)}$$

is then, essentially, only the one to one and onto correspondence of the right Kan extension $Ran_S(S)$.

$$r_o \quad \frac{(S \longrightarrow \mathcal{G}(\mathbb{T})) = (\mathbb{C} \xrightarrow{\overline{S}} \mathbb{A}^{\mathbb{T}}) \text{ "="} (TS \xRightarrow{s} S)}{(\mathbb{T} \longrightarrow \check{\mathcal{G}}(S)) = (T \Longrightarrow Ran_S(S))} \quad .$$

It is immediate from Propositions II.1.2 and II.1.5 that the arrow $\mathbb{T} \longrightarrow \check{\mathcal{G}}\,\mathcal{G}(\mathbb{T})$ in $\mathcal{M}(\mathbb{A})$, $\mathbb{T} \longrightarrow \mathbb{T}_{U^{\mathbb{T}}}$ is the equality. That is, the codensity V-monad of $U^{\mathbb{T}}$ is $\mathbb{T}$.

The arrow $S \longrightarrow \mathcal{G}\,\check{\mathcal{G}}(S)$ in $\mathcal{Z}_r(\mathbb{C}at, \mathbb{A})$, $\mathbb{C} \xrightarrow{\overline{S}} \mathbb{A}^{\mathbb{T}_S} \xrightarrow{U^{\mathbb{T}_S}} \mathbb{A}$, $S: \mathbb{C} \to \mathbb{A}$,

is $Ran_S(S)S \xRightarrow{r_o(id)} S$.

The V-functor $\overline{S}$ is called the <u>semantical comparison V-functor</u> of S. When S has a V-left adjoint we have:

• <u>Proposition II.1.6</u>

Given any V-functor $\mathbb{B} \xrightarrow{G} \mathbb{A}$ with a V-left adjoint $\mathbb{A} \xrightarrow{F} \mathbb{B}$; $(\epsilon, \eta): F \longrightarrow\!|_V G$, the semantic comparison V-functor of G, $\mathbb{B} \xrightarrow{\overline{G}} \mathbb{A}^{\mathbb{T}_G}$, is $GFG \xRightarrow{G\epsilon} G$ and is unique making the following two triangles commutative:

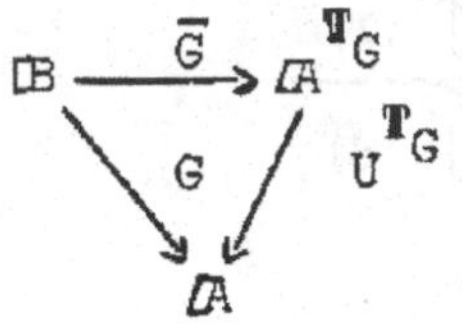

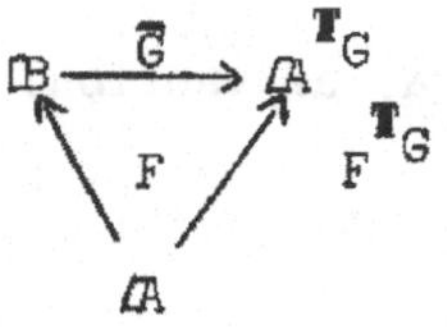

Proof:

From Proposition II.1.5 (and the rules for r_o given there) it follows that $\bar{G}$ is GFG $\overset{G\epsilon}{=\!=\!=\!\Longrightarrow}$ G. The composite $\bar{G}$F is then GFGF $\overset{G\epsilon F}{=\!=\!=\!\Longrightarrow}$ GF, and so it is equal to $F^{\mathbf{T}G}$ just by the definition of $F^{\mathbf{T}G}$. Finally, any other V-functor making 1) commutative has to be of the form GFG $\overset{g}{=\!=\!\Longrightarrow}$ G. If in addition it also makes 2) commutative, that is, gF = GϵF, it follows (again by the rules of r_o given in Proposition II.1.5) that $r_o(g) = r_o(G\epsilon)$, and so g = Gϵ. ∎

Section 2   Characterizations of Monadic V-functors

In this section we will prove the Beck triplability theorem, but first, we will isolate the particular case when the V-monad is idenpotent, a case that we will (explicitly) need later and for which it is not necessary to assume the existness of equalizers in $\mathbb{V}$.

A V-monad $\mathbf{T} = (T,\mu,\eta)$ is said to be idempotent if the multiplication TT $\overset{\mu}{=\!=\!\Longrightarrow}$ T is an isomorphism. It follows then that $\mu = (T\eta)^{-1}$, and since for any $\mathbf{T}$-algebra TA $\overset{a}{\longrightarrow}$ A,

$Ta \circ T\eta A = id$, we have $Ta = \mu A$. Then $id = \mu A \circ \eta TA = Ta \circ \eta TA = \eta A \circ a$ and so ηA is an isomorphism with inverse a. Referring ahead to Proposition II.4.5 (c) ⟹ d)), we know that in this case the two maps whose equalizer gives the V-structure for the category of algebras are equal, and so its equalizer is the identity map and therefore it always exists. Furthermore, $U^{\mathbb{T}}$ is a V-full-and-faithful V-functor.

<u>Proposition II.2.1</u>

Given any V-full-and-faithful V-functor $\mathbb{B} \xrightarrow{G} \mathbb{A}$ with a V-left adjoint $\mathbb{A} \xrightarrow{F} \mathbb{B}$, the semantical comparison V-functor $\mathbb{B} \xrightarrow{\overline{G}} \mathbb{A}^{\mathbb{T}}$ is a V-equivalence of V-categories (where we write $\mathbb{T}$ for $\mathbb{T}_G$).

<u>Proof</u>:

By Proposition 0.3 $\mathbb{T}$ is idempotent. Consider the V-functor $\mathbb{A}^{\mathbb{T}} \xrightarrow{FU^{\mathbb{T}}} \mathbb{B}$, we have:

$$FU^{\mathbb{T}}\overline{G} = FG \Longrightarrow id \text{ and } \overline{G}FU^{\mathbb{T}} = F^{\mathbb{T}}U^{\mathbb{T}} \Longrightarrow id.$$

The result follows then from Proposition 0.3. ∎

The above proposition can be generalized into a characterization of V-categories of algebras over a V-monad (in the ordinary set-based context the already classic Beck triplability theorem). Recall that given a V-functor $\mathbb{B} \xrightarrow{G} \mathbb{A}$, a pair of maps $A \overset{f}{\underset{g}{\rightrightarrows}} B$ in $\mathbb{B}$ is <u>G-contractible</u> if the pair of maps G(f), G(g), is part of a contractible coequalizer in $\mathbb{A}$. G detects V-coequalizers of G-contractible pairs if every G-contractible pair has a V-coequalizer. G preserves and reflects V-coequalizers of G-contractible pairs if for any G-contractible pair $A \overset{f}{\underset{g}{\rightrightarrows}} B$, a map $B \xrightarrow{h} C$ is a V-coequalizer of f and g if and only if $G(h)$ is a coequalizer (necessarily contractible) of $G(f)$ and $G(g)$. Finally, let us observe that for any V-monad $\mathbb{T}$, the forgetful V-functor $U^{\mathbb{T}}$ creates (that is, detects uniquely and preserves strictly) V-coequalizers of $U^{\mathbb{T}}$-contractible pairs. In particular, it detects, preserves and reflects. This can be seen easily as in the ordinary case and using the nine (3 x 3) lemma for equalizers in $\mathbb{V}$.

• Theorem II.2.1

Given any V-functor $\mathbb{B} \xrightarrow{G} \mathbb{A}$ with a V-left adjoint $\mathbb{A} \xrightarrow{F} \mathbb{B}$, G is monadic, that is, the semantical comparison V-functor $B \xrightarrow{\overline{G}} \mathbb{A}^{\mathbb{T}}$ is a V-equivalence of V-categories, if and only if G detects, preserves and reflects coequalizers of G-contractible pairs. It is a V-isomorphism of V-categories if and only if G creates coequalizers of G-contractible pairs.

Proof:

Let $id \xRightarrow{\eta} GF = U^{\mathbb{T}} F^{\mathbb{T}}$, $FG \xRightarrow{\epsilon} id$ and $F^{\mathbb{T}} U^{\mathbb{T}} \xRightarrow{\epsilon^{\mathbb{T}}} id$ (again we abbreviate $\mathbb{T}_G$ by $\mathbb{T}$)

As in proposition II.2.1 consider $\mathbb{A}^{\mathbb{T}} \xrightarrow{FU^{\mathbb{T}}} \mathbb{B}$,

$$(FU^{\mathbb{T}})\overline{G} = FG \xRightarrow{\epsilon} id \text{ and } \overline{G}(FU^{\mathbb{T}}) = F^{\mathbb{T}} U^{\mathbb{T}} \xRightarrow{\epsilon^{\mathbb{T}}} id.$$

It is easy to check that the following diagram is a contractible coequalizer of V-functors.

$$G(FU^{T})\overline{G}(FU^{T}) \underset{\underset{\eta\, GFU^{T}}{\longleftarrow}}{\overset{G\epsilon FU^{T}}{\underset{GFU^{T}\epsilon^{T}}{\rightrightarrows}}} G(FU^{T}) \underset{\eta\, U^{T}}{\overset{U^{T}\epsilon^{T}}{\rightleftarrows}} U^{T}$$

Therefore it necessarily is a pointwise contractible coequalizer, hence G detects it (pointwise) and so the pair of V-natural transformations ϵFU^{T} and $FU^{T}\epsilon^{T}$ has a (pointwise) coequalizer of V-functors. We define then the V-functor $A^{T} \xrightarrow{L} B$ as being that coequalizer. That is:

$$(FU^{T})\overline{G}(FU^{T}) \overset{\epsilon FU^{T}}{\underset{FU^{T}\epsilon^{T}}{\rightrightarrows}} FU^{T} \xrightarrow{g} L$$

Since G (pointwise) preserves this coequalizer we have $GL \approx U^{T}$ (V-natural isomorphisms). If G creates, then $GL = U^{T}$. We have obtained then $U^{T}\overline{G}L \approx U^{T}$ or $U^{T}\overline{G}L = U^{T}$, that is, for the lifted V-functors: $\overline{G}L \approx id$ or $\overline{G}L = id$.

Consider now the following diagram:

$$\begin{array}{ccccc}
(FU^{\mathbf{T}})\overline{G}(FU^{\mathbf{T}})\overline{G} & \underset{FU^{\mathbf{T}}\epsilon^{\mathbf{T}}\overline{G}}{\overset{\epsilon FU^{\mathbf{T}}\overline{G}}{\rightrightarrows}} & (FU^{\mathbf{T}})\overline{G} & \xrightarrow{g\overline{G}} & L\overline{G} \\
\| & & \| & & \\
FGFG & \underset{FG\epsilon}{\overset{\epsilon FG}{\rightrightarrows}} & FG & \xrightarrow{\epsilon} & id.
\end{array}$$

$G\epsilon FG$ and $GFG\epsilon$ is clearly a contractible pair with coequalizer $GFG \xrightarrow{G\epsilon} G$, which is then reflected (pointwise) by G and so the bottom row in the diagram is a coequalizer of V-functors. Clearly $\epsilon FU^{\mathbf{T}}\overline{G} = \epsilon FG$, and from the definition of $\epsilon^{\mathbf{T}}$ and Proposition II.1.6 it is also clear that $FU^{\mathbf{T}}\epsilon^{\mathbf{T}}\overline{G} = FG\epsilon$. So, since the top row is a coequalizer of V-functors by definition, we have $L\overline{G} \simeq id$ (V-natural isomorphisms). If G creates, we know $GL = U^{\mathbf{T}}$, and so $GL\overline{G} = U^{\mathbf{T}}\overline{G} = G$. Then, using again that G creates, we obtain $L\overline{G} = id$. ∎

Section 3 Clone of operations. V-codense and V-cogenerating V-functors

Given a tractable V-functor $\mathbb{C} \xrightarrow{S} \mathbb{A}$, since the representables preserve the codensity monad, for every $A \in \mathbb{A}$, $Ran_S \mathbb{A}(A,S(-))$ exists, and since any Kan extension with codomain V is pointwise, for any other $B \in \mathbb{A}$ the following end exists in $\mathbb{V}$:

$$\int_C \mathbb{V}(\mathbb{A}(B,SC), \mathbb{A}(A,SC))$$

(Recall that the above end is the formula given in Theorem I.4.2.) By the considerations made in I.5 (page 57) this is exactly $\mathbb{V}^{\mathbb{C}}(\mathbb{A}(B,S(-)), \mathbb{A}(A,S(-))$. So, if a V-functor S is tractable, for every $A, B \in \mathbb{A}$, the class of V-natural transformations between $\mathbb{A}(B,S(-))$ and $\mathbb{A}(A,S(-))$ is a set, and furthermore, it is the underlying set of an object of $\mathbb{V}$, namely, the end displayed above. There is no difficulty in checking that the class of objects of $\mathbb{A}$ together with the above end between them form a V-category, $\mathbb{K}_S$, <u>the clone of operations</u> of S, also called the Kleisli category of the codensity monad $\mathbb{T}_S$.

$$\mathbb{K}_S(A,B) = \mathbb{V}^{\mathbb{C}}(\mathbb{A}(B,S(-)), \mathbb{A}(A,S(-)) \quad .$$

The collection of maps (which is a V-natural family):

$$\mathbb{A}(A,B) \xrightarrow{\mathbb{A}(-,SC)} \mathbb{V}(\mathbb{A}(B,SC), \mathbb{A}(A,SC))$$

lifts into the end, providing a structure of V-functor to the identity map between objects:

$$\mathbb{A} \xrightarrow{F^S} \mathbb{K}_S \ , \quad F^S A = A$$

Since by definition (we use the fact that the representables preserve $Ran_S(S)$) $\mathbb{A}(A, Ran_S(S)(B)) = \mathbb{K}_S(A,B)$, the arrow:

$$\mathbb{K}_S \xrightarrow{U^S} \mathbb{A}, \quad U^S A = Ran_S(S)(A)$$

is actually a V-functor, V-right adjoint to F^S, and it is obvious that the V-monad induced by the pair $F^S \dashv_V U^S$ (i.e., the codensity monad of U^S) is $\mathbb{T}_S$.

Definition II.3.1

A V-functor $\mathbb{C} \xrightarrow{S} \mathbb{A}$ is V-codense if it is tractable and the unit of the codensity monad is an isomorphism.

Definition II.3.2

A V-functor $\mathbb{C} \xrightarrow{S} \mathbb{A}$ is V-cogenerating if it is tractable and the unit of the codensity monad is a pointwise V-monomorphism.

Clearly, any V-codense V-functor is V-cogenerating. If $\mathbb{C}$ is small we say that S is small V-codense and small V-cogenerating respectively. By Proposition 0.3, a V-functor $\mathbb{C} \xrightarrow{S} \mathbb{A}$ is V-cogenerating if and only if $\mathbb{A} \longrightarrow \mathbb{K}_S$ is V-faithful, which at the level of sets means that different morphisms $A \longrightarrow B$ in $\mathbb{A}$ determine different V-natural transformations $\mathbb{A}(B,S(-)) \Longrightarrow \mathbb{A}(A,S(-))$. If the base functor $\mathbb{V} \longrightarrow \mathbb{S}$ is not faithful this is no longer true for the induced natural transformations $\mathbb{A}_o(B, S(-)) \Longrightarrow \mathbb{A}_o(B,S(-))$, and so, a V-cogenerating V-functor is not necessarily cogenerating. Similarly, $\mathbb{C} \xrightarrow{S} \mathbb{A}$ is V-codense if and only if $\mathbb{A} \longrightarrow \mathbb{K}_S$ is V-full-and-faithful (that is, $\mathbb{K}_S = \mathbb{A}$), which at the level of sets means that any V-natural transformation

$\mathbb{A}(B,S(-)) \Longrightarrow \mathbb{A}(A,S(-))$ is necessarily and uniquely determined by a morphism $A \longrightarrow B$. For the same reasons as before, a V-codense V-functor will in general fail to be codense.

When $\mathbb{C}$ is the V-category $\mathbb{1}$ ($\mathbb{1}$ has only one object $1 \in \mathbb{1}$, and $\mathbb{1}(1,1) = I$) a V-functor $\mathbb{1} \xrightarrow{S} \mathbb{A}$ is completely characterized by an object $S \in \mathbb{A}$ and viceversa. In this case we say that S is a <u>V-cogenerator</u> or a <u>V-codense V-cogenerator</u> of $\mathbb{A}$. It follows from Theorem I.4.2 that the codensity monad is given by the pair of V-adjoint functors $\mathbb{A} \xrightarrow{\mathbb{A}(-,S)} \mathbb{V}^{op}$, $\mathbb{V}^{op} \xrightarrow{\mathbb{A}(-,S)} \mathbb{A}$, hence S is a V-cogenerator if and only if the representable V-functor $\mathbb{A}(-, S)$ is V-faithful, it is a V-codense V-cogenerator if $\mathbb{A}(-, S)$ is V-full-and-faithful.

Considering the dual concept; the object I is always a V-dense V-generator of $\mathbb{V}$.

We now introduce a notion that will simplify the exposition of our next statement and of some other arguments in what remains of the paper. Given a functor $\Gamma \xrightarrow{\Gamma} \mathbb{A}$ from any category Γ into a V-category $\mathbb{A}$, suppose the $\text{V-}\varprojlim_{\lambda} \Gamma_\lambda$ exists in $\mathbb{A}$; then we say that a V-functor $\mathbb{A} \xrightarrow{H} \mathbb{A}'$ <u>conserves</u> the $\text{V-}\varprojlim_{\lambda} \Gamma_\lambda$ if $\text{V-}\varprojlim_{\lambda} H\Gamma_\lambda$ also exists in $\mathbb{A}'$. (for example, any V-functor which preserves V-limits conserves any V-limit that might exist). Similarly there is the notion of conservation of ends, cotensors and right Kan extensions.

Proposition II.3.1

Given any cotensored V-category $\mathbb{A}$ and a V-monad $\mathbb{T}$ in $\mathbb{A}$, let $\mathbb{B}$ be the V-full sub-category whose objects are those objects A of $\mathbb{A}$ for which the unit of the V-monad $A \xrightarrow{\eta A} TA$ is a V-monomorphism. Then, $\mathbb{B}$ is closed under cotensors and under all the V-limits and ends which are conserved by T. Furthermore, $\mathbb{B}$ is also closed under V-sub-objects.

Proof:

Let $\mathbb{C} \xrightarrow{\Gamma} \mathbb{B}$ be any functor and suppose V-$\varprojlim_{\lambda} \Gamma_\lambda$ exists in $\mathbb{A}$ and is conserved by T. Consider the diagram:

$$\begin{array}{ccc}
\text{V-}\varprojlim_{\lambda}\Gamma_\lambda & \xrightarrow{\ \eta\,\text{V-}\varprojlim_{\lambda}\Gamma_\lambda\ } & T\,\text{V-}\varprojlim_{\lambda}\Gamma_\lambda \\
 & {\scriptstyle \text{V-}\varprojlim_{\lambda}\eta\Gamma_\lambda}\searrow \qquad \swarrow{\scriptstyle z} & \\
 & \text{V-}\varprojlim_{\lambda} T\,\Gamma_\lambda &
\end{array}$$

which can easily be seen to be commutative. From

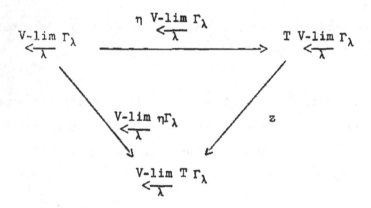

Proposition I.1.1 we know that $\underset{\overleftarrow{\lambda}}{\text{V-lim}}\ \eta\Gamma_\lambda$ is a V-monomorphism, so η $\underset{\overleftarrow{\lambda}}{\text{V-lim}}\ \Gamma$ is a V-monomorphism. For ends we proceed similarly; the result follows from the corresponding Proposition I.3.1. For cotensors: let $V \in \mathbb{V}$ and $B \in \mathbb{B}$, consider the diagram:

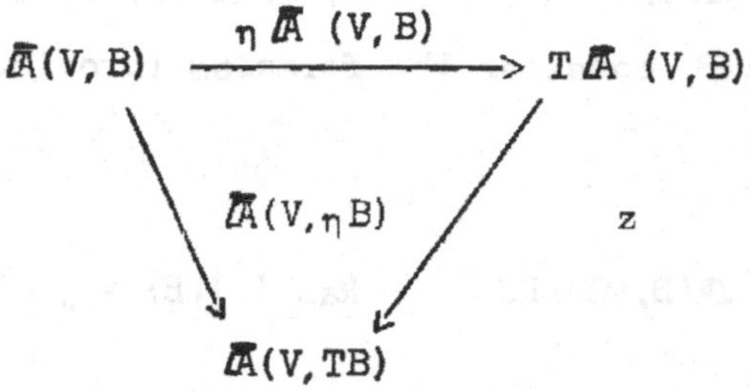

It is easy to see that $\mathbb{A}(V,-)$ sends V-monomorphisms into V-monomorphisms, therefore $\mathbb{A}(V,\eta B)$ is a V-monomorphism, and so, the fact that $\mathbb{A}(V,B) \in \mathbb{B}$ will follow from the commutativity of the diagram, which offers no difficulty going to the other side of the adjointness by σ_o and recalling the definition of z ((1) page 24). Finally, if $A \longrightarrow B$ is a V-monomorphism, it is trivial that if ηB is a V-monomorphism so is ηA, that is, $\mathbb{B}$ is closed under V-sub-objects. ∎

<u>Proposition II.3.2</u>

Let $\mathbb{A}$ be any cotensored V-category and $\mathbb{C} \xrightarrow{R} \mathbb{A}$ any tractable V-functor. Let $\mathbb{B}$ be as in the previous proposition (with respect to the codensity V-monad $\mathbb{T}_R$). Then, R factors through $\mathbb{B}$, $\begin{array}{ccc} \mathbb{C} & \xrightarrow{R} & \mathbb{A} \\ & {\scriptstyle S}\searrow & \nearrow \\ & & \mathbb{B} \end{array}$ and the V-functor $\mathbb{C} \xrightarrow{S} \mathbb{B}$ is V-cogenerating.

<u>Proof</u>:

For every $C \in \mathbb{C}$, RC has a structure of $\mathbb{T}_R$-algebra, in particular, ηRC splits and so $RC \in \mathbb{B}$, that is, R factors through $\mathbb{B}$. Given any $B \in \mathbb{B}$, consider the formulas (provided by Theorem I.4.2):

$$\mathrm{Ran}_R(R)(B) = \int_C \bar{\bar{\mathbb{A}}}(\mathbb{A}(B,RC),RC) \qquad \mathrm{Ran}_S(S)(B) = \int_C \bar{\bar{\mathbb{B}}}(\mathbb{B}(B,SC),SC)$$

For the same reasons as for R, $\mathrm{Ran}_R(R)$ also factors through $\mathbb{B}$. The left end, which exists by assumption, belongs then to $\mathbb{B}$, and so it is an end in $\mathbb{B}$. Noticing that the previous proposition shows that $\mathbb{B}$ is also cotensored and $\bar{\bar{\mathbb{A}}}(V,B) = \bar{\bar{\mathbb{B}}}(V,B)$ for any $V \in \mathbb{V}$, $B \in \mathbb{B}$, we deduce that the right end also exists, and so $\mathrm{Ran}_S(S)$ exists pointwise. That is, S is tractable. Furthermore, for any $B \in \mathbb{B}$, $\mathrm{Ran}_R(R)(B) = \mathrm{Ran}_S(S)(B)$, hence, by definition of $\mathbb{B}$, S is V-cogenerating. ∎

Observe that any sub-object in $\mathbb{A}$ of a V-limit and/or end (conserved by $T = \mathrm{Ran}_R(R)$) of cotensors of objects of the form RC, $C \in \mathbb{C}$, belongs to $\mathbb{B}$. Hence, denoting by $\mathbb{B}'$ the V-full sub-category so defined, we have $\mathbb{B}' \subset \mathbb{B}$. If for every $B \in \mathbb{B}$,

$Ran_R(R)$ conserves the end in the formula for $Ran_R(R)(B)$, we have $\mathbb{B}' = \mathbb{B}$. In any case the following holds:

Remark II.3.1

Let $\mathbb{A}$ be any cotensored V-category and $\mathbb{C} \xrightarrow{R} \mathbb{A}$ any tractable V-functor. If every object of $\mathbb{A}$ is a V-sub-object of a V-limit or end (conserved by $Ran_R(R)$) of cotensors of objects of the form RC $C \in \mathbb{C}$, then R is V-cogenerating (the converse holds if $Ran_R(R)$ conserves the end $Ran_R(R)(A)$ for every $A \in \mathbb{A}$).

Proof:

Just notice that in this case $\mathbb{A} \subset \mathbb{B}$ and so $\mathbb{A} = \mathbb{B}$ ($\mathbb{B}$ as in the previous proposition). ∎

Let us remark that conveniently stated versions of the last two Propositions and Remark could, (by means of more elaborate arguments) have been proved without the restriction that $\mathbb{A}$ be cotensored.

Finally, let us also remark that given a V-functor $\mathbb{C} \xrightarrow{R} \mathbb{A}$, the fact that every object of $\mathbb{A}$ is a V-limit (or end) of cotensors of objects of the form RC, $C \in \mathbb{C}$ is not at all enough for R to be V-codense, as can be seen, for example, with the inclusion of finite sets into sets (at least if we assume the Continuum Hypotheses). However, in the unusual case in which R, besides being V-full-and-faithful, is such that $Ran_R(R)$ preserves V-limits, ends and cotensors, this weaker (non canonical) V-codensity implies V-codensity as defined in this paper (that is, $Ran_R(R) \approx$ id).

Section 4. Additional Properties

Proposition II.4.1

Given any V-functor $\mathbb{C} \xrightarrow{S} \mathbb{B}$ which admits a codensity V-monad and a V-functor $\mathbb{B} \xrightarrow{G} \mathbb{A}$ which preserves $Ran_S(S)$ and such that $\mathbb{C} \xrightarrow{R} \mathbb{A}$, $R = GS$, also admits a codensity V-monad, then, there is a V-natural transformation $Ran_R(R)G \Longrightarrow G\, Ran_S(S)$ making the following diagram commutative:

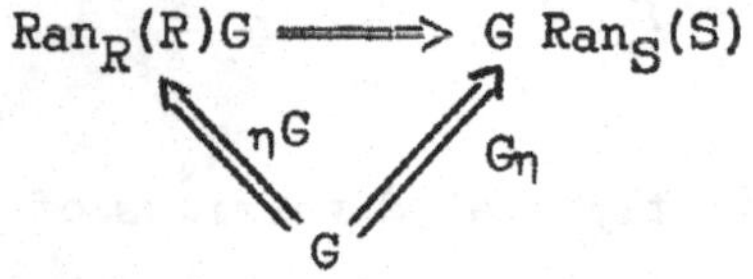

(we write η for the units of both codensity V-monads)

Proof:

Define $Ran_R(R)G \Longrightarrow G\, Ran_S(S)$ as the composite $z^{-1} \circ \epsilon \circ \theta G$ in the diagram below:

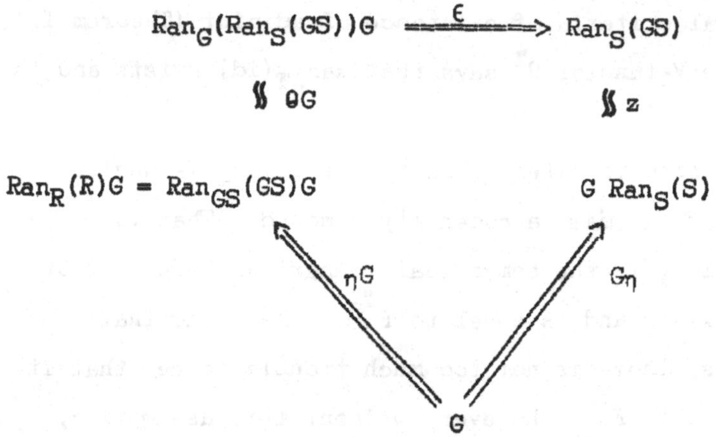

The commutativity of this diagram is equivalent by r_o to that of the diagram.

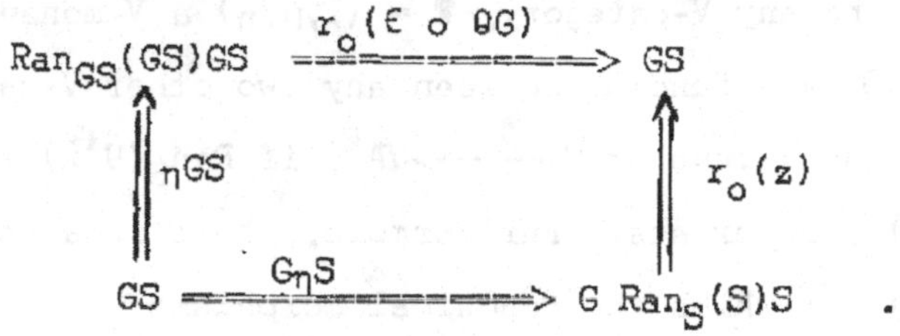

But the commutative diagrams (1) on pages 42 and 43 mean that $r_o(z) = G\epsilon$ and $r_o(z) = G\epsilon$ and $r_o(\epsilon \circ \theta G) = \epsilon$, so, from diagram (1) in page 47 it follows that both paths in the above diagram are the identity and therefore equal. ∎

The formal criteria of existence of adjoint (Theorem I.4.1) applied to the V-functor $U^{\mathbb{T}}$ says that $\mathrm{Ran}_{U^{\mathbb{T}}}(\mathrm{id})$ exists and is equal to $F^{\mathbb{T}}$.

We would like to extend this result to any V-functor $\mathbb{C} \xrightarrow{S} \mathbb{A}$ that admits a codensity V-monad. That is, (recall that $\mathrm{id}_{\mathbb{A}^{\mathbb{T}}}$ is the semantical comparison V-functor of $U^{\mathbb{T}}$) $\mathrm{Ran}_S(\overline{S})$ exists and is equal to $F^{\mathbb{T}}S$. Assuming that $\mathrm{Ran}_S(\overline{S})$ exists, there is not too much trouble to see that it has to be equal to $F^{\mathbb{T}}S$; however, without this assumption, the only path left is to prove that $F^{\mathbb{T}}S$ is the right Kan extension of $\overline{S}$ along S. This is just a particular case of the expected property that $U^{\mathbb{T}}$ creates right Kan extensions. Explicitly

• Proposition II.4.2

Let $\mathbb{A}$ be any V-category, $\mathbb{T} = (T,\mu,\eta)$ a V-monad in $\mathbb{A}$ and $\mathbb{C} \xrightarrow{S} \mathbb{B}$ a V-functor between any two other V-categories. Then, given any V-functor $\mathbb{C} \xrightarrow{\overline{R}} \mathbb{A}^{\mathbb{T}}$, if $\mathrm{Ran}_S(U^{\mathbb{T}}\overline{R})$ exists, then $\mathrm{Ran}_S(\overline{R})$ also exists. Furthermore, there is a unique such $\mathrm{Ran}_S(\overline{R})$ for which the canonical morphisms $U^{\mathbb{T}}\,\mathrm{Ran}_S(\overline{R}) \overset{z}{\Longrightarrow} \mathrm{Ran}_S(U^{\mathbb{T}}\overline{R})$ is the equality. In other words, there is a unique $\mathbb{T}$-action on $\mathrm{Ran}_S(R)$ $(R = U^{\mathbb{T}}\overline{R})$ which makes the lifted V-functor $\overline{\mathrm{Ran}_S(R)}$ be a $\mathrm{Ran}_S(\overline{R})$ such that z is the equality. The diagramatic configuration is:

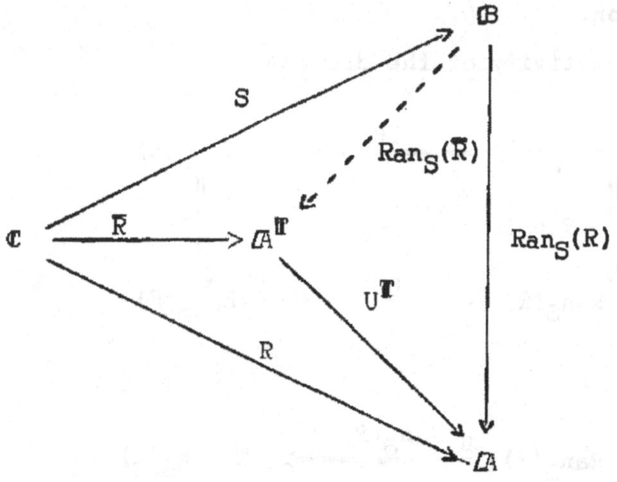

$$U^{\mathbb{T}} \, \text{Ran}_S(\bar{R}) = \text{Ran}_S(R) \qquad .$$

<u>Proof</u>:

For any V-functor $\mathbb{B} \xrightarrow{H} \mathbb{A}$ we have

$$r_o \; \frac{H \overset{\varphi}{\Longrightarrow} \text{Ran}_S(R)}{HS \overset{\psi}{\Longrightarrow} R} \qquad .$$

The V-functor $\bar{R}$ is $TR \overset{r}{\Longrightarrow} R$ (recall the considerations made in page 74).

Define $$r_o \; \frac{T\,\text{Ran}_S(R) \overset{\xi}{\Longrightarrow} \text{Ran}_S(R)}{T\,\text{Ran}_S(R)S \overset{T\epsilon}{\Longrightarrow} TR \overset{r}{\Longrightarrow} R} \qquad ,$$

where $\epsilon = r_o(\text{id})$.

ξ is a $\mathbb{T}$-Action.

The commutativity of the diagrams

$$\begin{array}{ccc} TT\,\mathrm{Ran}_S(R) & \xrightarrow{T\xi} & T\,\mathrm{Ran}_S(R) \\ \Big\Downarrow{\scriptstyle \mu\,\mathrm{Ran}_S(R)} & & \Big\Downarrow{\scriptstyle \xi} \\ T\,\mathrm{Ran}_S(R) & \xrightarrow{\xi} & \mathrm{Ran}_S(R) \end{array}$$

and

$$\begin{array}{ccc} \mathrm{Ran}_S(R) & \xrightarrow{\eta\,\mathrm{Ran}_S(R)} & T\,\mathrm{Ran}_S(R) \\ & {\scriptstyle id}\searrow & \Big\Downarrow{\scriptstyle \xi} \\ & & \mathrm{Ran}_S(R) \end{array}$$

are equivalent by r_o to that of the exterior of the diagrams.

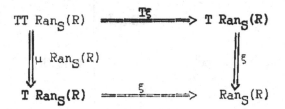

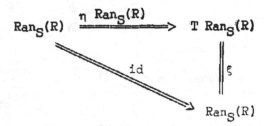

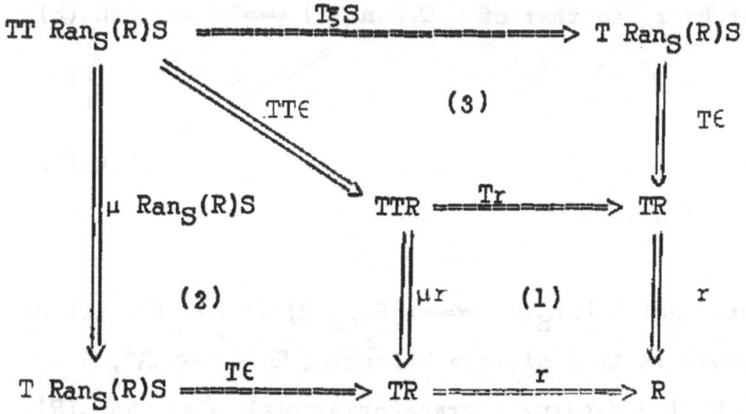

and

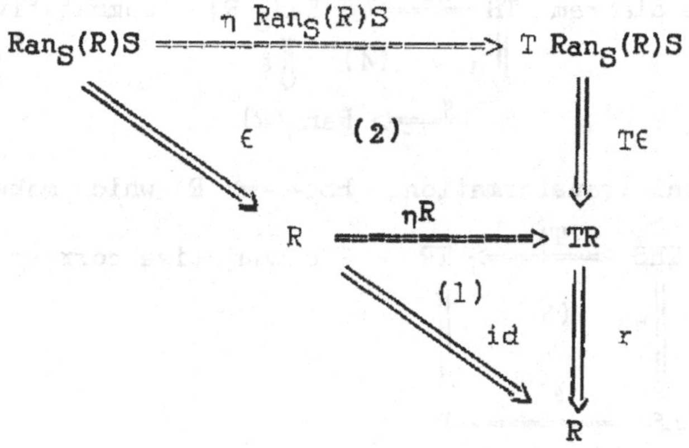

Diagrams (1) commute because r is a **T**-action and diagrams (2) commute by naturality. The commutativity of diagram (3) follows from that of the diagram

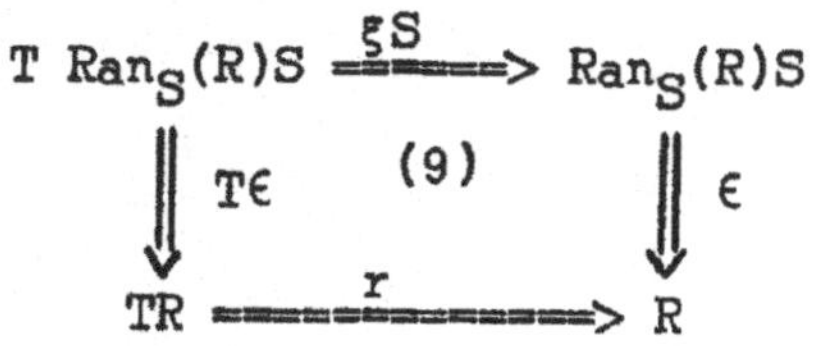

which is

equivalent by r_o to that of

$$\begin{array}{ccc} T\,Ran_S(R) & \overset{\xi}{\Longrightarrow} & Ran_S(R) \\ & \underset{\xi}{\searrow} & \Big\Downarrow id \\ & & Ran_S(R) \end{array} .$$

So ξ is a $\mathbb{T}$ action.

To see that $T\,Ran_S(R) \overset{\xi}{\Longrightarrow} Ran_S(R)$ is $Ran_S(\bar{R})$ all we have to prove is that given a V-functor $\mathbb{B} \xrightarrow{\bar{H}} \mathbb{A}^{\mathbb{T}}$, $TH \overset{h}{\Longrightarrow} H$, the V-natural transformations $H \overset{\varphi}{\Longrightarrow} Ran_S(R)$ which make the diagram

$$\begin{array}{ccc} TH & \overset{T\varphi}{\Longrightarrow} & T\,Ran_S(R) \\ \Big\Downarrow h & (4) & \Big\Downarrow \xi \\ H & \overset{\varphi}{\Longrightarrow} & Ran_S(R) \end{array}$$

commutative and the V-natural transformations $HS \overset{\psi}{\Longrightarrow} R$ which make the diagram

$$\begin{array}{ccc} THS & \overset{T\psi}{\Longrightarrow} & TR \\ \Big\Downarrow hS & (5) & \Big\Downarrow r \\ HS & \overset{\psi}{\Longrightarrow} & R \end{array}$$

commutative correspond each other under r_o.

Let φ be any V-natural transformation and consider the diagram:

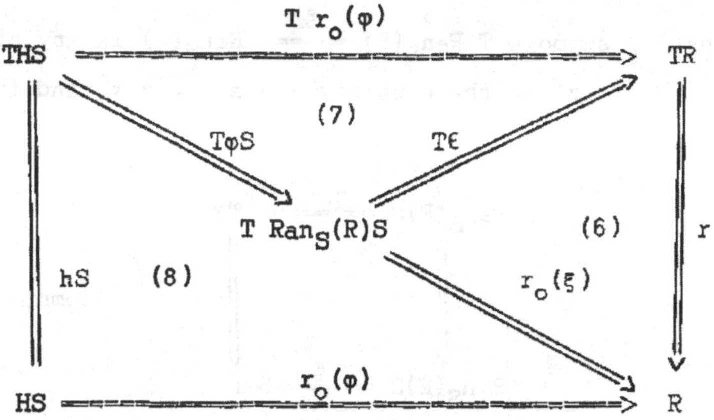

(6) is just the definition of ξ and diagram (7) commutes because it is the V-functor T applied to a diagram whose commutativity is equivalent by r_o by the commutativity of

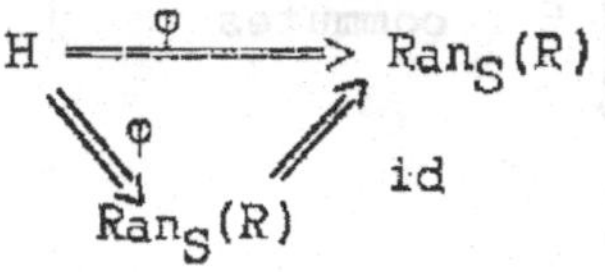

.

Since the commutativity of diagram (4) is equivalent by r_o to that of diagram (8), φ makes diagram (4) commutative if and only if $r_o(\varphi)$ makes diagram (5) commutative.

Observe that since the one to one and onto correspondence $\overline{r}_o$ of the $Ran_S(\overline{R})$ above defined is just r_o, the V-natural transformation $Ran_S(\overline{R})S \xrightarrow{\overline{\epsilon}} \overline{R}$ is just ϵ, that is, $U^{T}\overline{\epsilon} = \epsilon$. This means exactly that z is the equality (see (1) page 42).

Finally, suppose $T\,Ran_S(R) \overset{\xi'}{\Longrightarrow} Ran_S(R)$ is any other $Ran_S(\overline{R})$. That z be the equality means $U^T\overline{\xi}' = \epsilon$ and this implies that the

$$\begin{array}{ccc} T\,Ran_S(R)S & \overset{T\epsilon}{\Longrightarrow} & TR \\ \Big\Downarrow{\scriptstyle \xi' S} & & \Big\Downarrow{\scriptstyle r} \\ Ran_S(R)S & \overset{\xi}{\Longrightarrow} & R \end{array}$$

diagram commutes.

From this and diagram (9) it follows that the

$$\begin{array}{ccc} T\,Ran_S(R)S & \overset{\xi S}{\Longrightarrow} & Ran_S(R)S \\ \Big\Downarrow{\scriptstyle \xi' S} & & \Big\Downarrow{\scriptstyle \epsilon} \\ Ran_S(R)S & \overset{\xi}{\Longrightarrow} & R \end{array}$$

diagram commutes

But the commutativity of this latter diagram is equivalent by r_o to that of

$$\begin{array}{ccc} T\,Ran_S(R) & \overset{\xi}{\Longrightarrow} & Ran_S(R) \\ \Big\Downarrow{\scriptstyle \xi'} & & \Big\Downarrow{\scriptstyle id} \\ Ran_S(R) & \overset{id}{\Longrightarrow} & Ran_S(R) \end{array}$$

, that is, $\xi = \xi'$. ∎

• <u>Proposition II.4.3</u>

In the situation (or data) of the previous proposition, if $Ran_S(R)$ is preserved by all the representables, then

$\mathrm{Ran}_S(\bar{R})$ is also preserved by all the representables.

<u>Proof</u>:

We can assume that for every $A \in \mathbb{A}$ $\mathrm{Ran}_S(\mathbb{A}(A,R))$ is $\mathbb{A}(A, \mathrm{Ran}_S(R))$ with the universal V-natural transformation $r_o(\mathrm{id})$ given by $\mathbb{A}(\square, r_o(\mathrm{id})) = \mathbb{A}(\square, \epsilon)$. It follows then that given any V-natural transformation $H \overset{\theta}{\Longrightarrow} \mathrm{Ran}_S(R)$, $r_o(\mathbb{A}(\square,\theta)) = \mathbb{A}(\square, r_o(\theta))$.

Let $\bar{A} \in \mathbb{A}^{\mathbb{T}}$, $\bar{A} = (TA \xrightarrow{a} A)$. For any V-functor $\mathbb{B} \xrightarrow{F} \mathbb{V}$, consider, with the r_o given by the hypotheses:

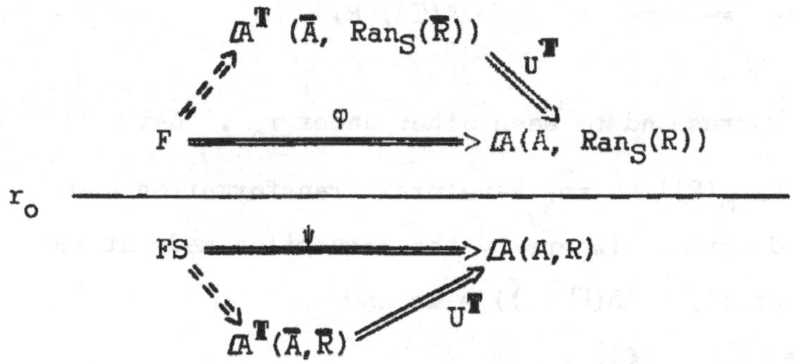

In order to see that $\mathrm{Ran}_S(\mathbb{A}^{\mathbb{T}}(\bar{A},\bar{R}))$ is $\mathbb{A}^{\mathbb{T}}(\bar{A}, \mathrm{Ran}_S(\bar{R}))$ it will be enough to prove that V-natural transformations admitting a factorization through $U^{\mathbb{T}}$ correspond each other under r_o .

By definition of $U^{\mathbb{T}}$, and because $U^{\mathbb{T}}$ being V-faithful reflects V-naturality, it is equivalent to prove that V-natural transformations φ which make the diagram:

$$\begin{array}{ccc}
F \overset{\varphi}{\Longrightarrow} \mathcal{A}(A, \mathrm{Ran}_S(R)) & \overset{T}{\Longrightarrow} & \mathcal{A}(TA, T\,\mathrm{Ran}_S(R)) \\
\Big\Downarrow \varphi & (1) & \Big\Downarrow \mathcal{A}(\square, \xi) \\
\mathcal{A}(A, \mathrm{Ran}_S(R)) & \overset{\mathcal{A}(a, \square)}{\Longrightarrow} & \mathcal{A}(TA, \mathrm{Ran}_S(R))
\end{array}$$

commutative and V-natural transformations ψ which make the diagram:

$$\begin{array}{ccc}
FS \overset{\psi}{\Longrightarrow} \mathcal{A}(A,R) & \overset{T}{\Longrightarrow} & A(TA, TR) \\
\Big\Downarrow \psi & (2) & \Big\Downarrow \mathcal{A}(\square, r) \\
\mathcal{A}(A,R) & \overset{\mathcal{A}(a, \square)}{\Longrightarrow} & \mathcal{A}(TA, R)
\end{array}$$

commutative, correspond to each other under r_o . Let $F \overset{\varphi}{\Longrightarrow} \mathcal{A}(A, \mathrm{Ran}_S(R))$ be any V-natural transformation and consider the diagram: (where by the assumption made at the start of the proof, $r_o(\mathcal{A}(\square, \epsilon)) = \mathrm{id}$ and $r_o(\mathcal{A}(\square, \xi)) = \mathcal{A}(\square, r_o(\xi))$.

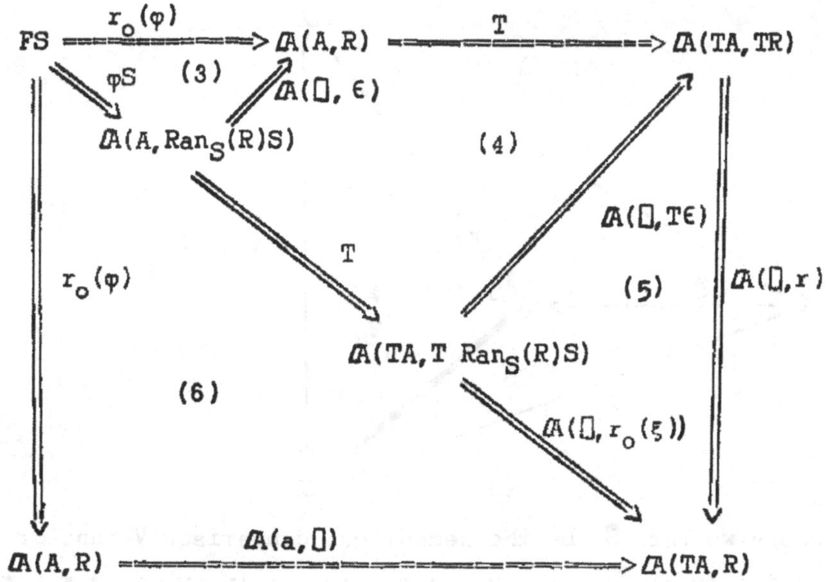

Using r_o to go to the other side of the adjointness, it is immediate that diagram (3) commutes. Diagram (4) commutes by naturality of T as a transformation of bifunctors, and diagram (5) because it is $\mathbb{A}(\square,-)$ applied to the diagram (6) of Proposition II.4.2. Since the commutativity of diagram (1) is equivalent by r_o to that of diagram (6), φ makes diagram (1) commutative if and only if $r_o(\varphi)$ makes diagram (2) commutative. ∎

Given any V-functor $\mathbb{C} \xrightarrow{S} \mathbb{A}$ admitting a codensity V-monad, we obtain a situation like the one in Proposition II.4.2 whose diagramatic configuration is:

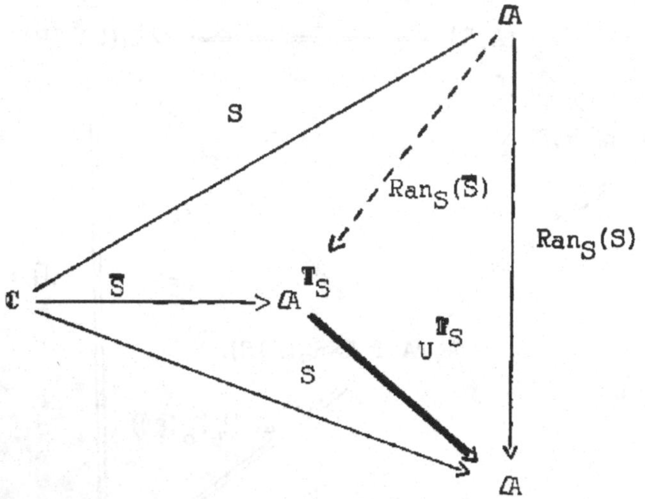

and where we let $\bar{S}$ be the semantical comparison V-functor of S. $\mathrm{Ran}_S(S)$ is now the V-endofunctor of the V-monad $\mathbb{T} = \mathbb{T}_S$, and the action ξ is (just by definition) the multiplication μ of $\mathbb{T}_S$. So $\mathrm{Ran}_S(\bar{S})$ is $F^{\mathbb{T}_S}$ by definition of $F^{\mathbb{T}_S}$.

• <u>Proposition II.4.4</u>

Given any V-functor $\mathbb{C} \xrightarrow{S} \mathbb{A}$ admitting a codensity V-monad, $\mathrm{Ran}_S(\bar{S})$ exists and $F^{\mathbb{T}_S} = \mathrm{Ran}_S(S)$. Furthermore, if S is tractable, then $\mathrm{Ran}_S(\bar{S})$ is preserved by all representables. ▮

Given any V-category $\mathbb{A}$ and a V-monad $\mathbb{T} = (T,\mu,\eta)$ in $\mathbb{A}$, the objects of $\mathbb{A}$ in which η is an isomorphism can be characterized in several ways.

• Proposition II.4.5

For an object $A \in \mathbb{A}$; the following four properties are equivalent. Furthermore, in any of the four cases, μA is necessarily an isomorphism and $\bar{A}$ is unique such that $U^{\mathbb{T}}(\bar{A}) = A$.

a) A has a $\mathbb{T}$-algebra structure $\bar{A} = (TA \xrightarrow{a} A)$ such that $A \xrightarrow{\eta A} TA$ is a morphism of algebras, $\bar{A} \longrightarrow F^{\mathbb{T}}A$.

b) $A \xrightarrow{\eta A} TA$ is an isomorphism.

c) $A \xrightarrow{\eta A} TA$ is an isomorphism and $(\eta A)^{-1}$ is a $\mathbb{T}$-algebra structure.

d) A has a $\mathbb{T}$-algebra structure $\bar{A} = (TA \xrightarrow{a} A)$ such that for any other $\bar{B} \in \mathbb{A}^{\mathbb{T}}$, $\bar{B} = (TB \xrightarrow{b} B)$, $U^{\mathbb{T}}_{\bar{A}\bar{B}}$ is an isomorphism, i.e., $\mathbb{A}^{\mathbb{T}}(\bar{A},\bar{B}) \approx \mathbb{A}(A,B)$, that is, $\mathbb{A}^{\mathbb{T}}(\bar{A},-) \xrightarrow{U^{\mathbb{T}}} \mathbb{A}(A,U^{\mathbb{T}}(-))$ is an isomorphism.

Proof:

(a) ⟹ b)).

Consider the diagram expressing the fact that ηA is a morphism of algebras:

$$(1)\qquad \begin{array}{ccc} TA & \xrightarrow{T\eta A} & TTA \\ \downarrow a & & \downarrow \mu A \\ A & \xrightarrow{\eta A} & TA \end{array} \quad . \qquad \text{Then } \eta A \circ a = \text{id}.$$

So a is an inverse for ηA.

(b) ===> c))

Just write down the diagrams

(c) ===> d))

The exterior of the following diagram commutes:

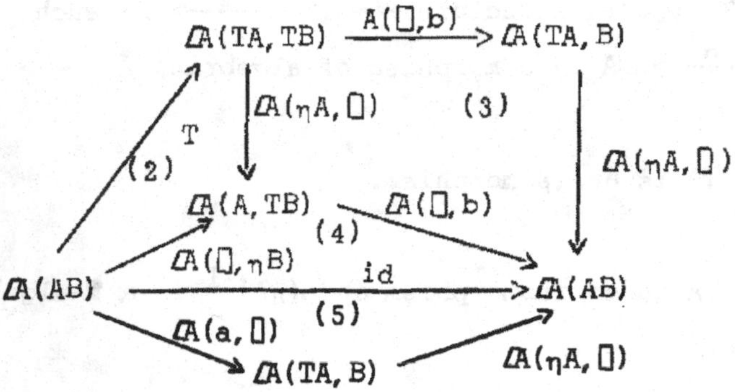

(2) commutes because η is V-natural, (3) because $\mathcal{A}(-,-)$ is a bifunctor and (4) and (5) because b and a are T-algebra structures.

Then; since $A(\eta A, \square)$ is an isomorphism, the following diagram commute:

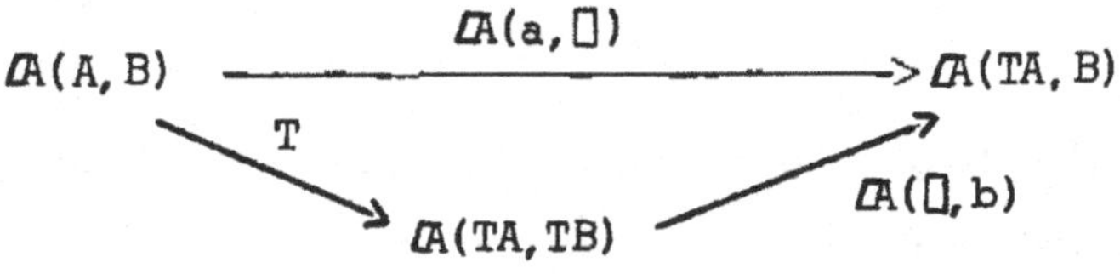

So, by definition, $\mathcal{A}^{\mathbb{T}}(\bar{A},\bar{B}) \approx \mathcal{A}(AB)$ and $U^{\mathbb{T}}$ is an isomorphism.

d) $\Longrightarrow$ a)

$$\text{Since } \mathbb{A}^{\mathbb{T}}{}_{o}(\bar{A}, F^{\mathbb{T}}A) = \mathbb{A}_{o}(A, TA)$$

Finally; from diagram (1) we see that μA is an isomorphism. The uniqueness of $\bar{A}$ is clear. ∎

• Proposition II.4.6

Given any tractable V-functor $\mathbb{C} \xrightarrow{S} \mathbb{A}$, if $\bar{A} \in \mathbb{A}^{\mathbb{T}_S}$ is such that $\mathbb{A}^{\mathbb{T}_S}(\bar{A}, \bar{S}) \approx \mathbb{A}(A, S)$ ($\bar{S}$ the semantical comparison V-functor), then ηA is an isomorphism or equivalently (By Proposition II.4.5) $\mathbb{A}^{\mathbb{T}_S}(\bar{A}, -) \approx \mathbb{A}(A, -)$.

Proof:

Using Proposition II.4.4 we have:

$\mathbb{A}^{\mathbb{T}_S}(\bar{A}, F^{\mathbb{T}}) = \mathbb{A}^{\mathbb{T}_S}(\bar{A}, \mathrm{Ran}_S(\bar{S})) \approx \mathrm{Ran}_S(\mathbb{A}^{\mathbb{T}_S}(\bar{A}, \bar{S})) \approx \mathrm{Ran}_S(\mathbb{A}(A, S)) \approx$ $\approx \mathbb{A}(A, \mathrm{Ran}_S(S))$. Then, using this isomorphism at the level of sets it follows that ηA is a morphism of $\mathbb{T}_S$-algebras, hence by Proposition II.4.5 we are done. ∎

The creativity property of the forgetful functor of set-based category theory with respect to limits holds in the V-context with respect to the three different notions related with completeness. That is, the forgetful V-functor creates any cotensors, V-limits or ends that might exist. Explicitly, let $\mathbb{B}$ be any V-category and $\mathbb{T}$ a V-monad in $\mathbb{B}$. Then:

• <u>Proposition II.4.7</u>

Given $\overline{B} \in \mathbb{B}^{\mathbf{T}}$, $\overline{B} = (TB \xrightarrow{b} B)$ and $V \in \mathbb{V}$, if $\overline{\mathbb{B}}(V,B)$ exists, then $\overline{\mathbb{B}^{\mathbf{T}}}(V,\overline{B})$ also exists. Furthermore, there is a unique one such that the canonical morphism $U^{\mathbf{T}}\,\overline{\mathbb{B}^{\mathbf{T}}}\,(V,\overline{B}) \xrightarrow{z} \overline{\mathbb{B}}\,(V,U^{\mathbf{T}}\overline{B})$ is the equality. In other words, there is a unique structure of $\mathbf{T}$-algebra on $\overline{\mathbb{B}}(V,B)$ which is a cotensor $\overline{\mathbb{B}^{\mathbf{T}}}\,(V,\overline{B})$ and which makes z the equality.

<u>Proof</u>:

A map $T\overline{\mathbb{B}}(V,B) \xrightarrow{\xi} \overline{\mathbb{B}}(V,B)$ is given by

$$\overline{\mathbb{B}}(V,B) \xrightarrow{id} \overline{\mathbb{B}}(V,B)$$

σ_o ————————————————

$$V \longrightarrow \mathbb{B}(\overline{\mathbb{B}}(V,B),B) \xrightarrow{T} \mathbb{B}(T\overline{\mathbb{B}}(V,B),TB) \xrightarrow{\mathbb{B}(\square,b)} \mathbb{B}(T\overline{\mathbb{B}}(V,B),B)$$

σ_o ————————————————————————————————

$$T\,\overline{\mathbb{B}}(V,B) \xrightarrow{\xi} \overline{\mathbb{B}}(V,B)$$

By the representation theorem, ξ makes the following diagram commutative for every $A \in \mathbb{B}$.

$$\begin{array}{ccc}
\mathbb{B}(A,\bar{\mathbb{B}}(V,B)) & \xrightarrow{\quad\sigma\quad} & \mathbb{V}(V,\mathbb{B}(A,B)) \\
\Big\downarrow {\scriptstyle T} & & \Big\downarrow {\scriptstyle \mathbb{V}(\square,T)} \\
\mathbb{B}(TA,T\bar{\mathbb{B}}(V,B)) & (1) & \mathbb{V}(V,\mathbb{B}(TA,TB)) \\
\Big\downarrow {\scriptstyle B(\square,\xi)} & & \Big\downarrow {\scriptstyle \mathbb{V}(\square,\mathbb{B}(\square,b))} \\
\mathbb{B}(TA,\bar{\mathbb{B}}(V,B)) & \xrightarrow{\quad\sigma\quad} & \mathbb{V}(V,\mathbb{B}(TA,B))
\end{array}$$

In order to see that ξ is a $\mathbb{T}$-algebra structure we do as follows:

The commutativity of diagram

$$\begin{array}{ccc}
TT\bar{\mathbb{B}}(V,B) & \xrightarrow{\quad T\xi\quad} & T\bar{\mathbb{B}}(V,B) \\
\Big\downarrow {\scriptstyle \mu\bar{B}(V,B)} & (2) & \Big\downarrow {\scriptstyle \xi} \\
T\bar{\mathbb{B}}(V,B) & \xrightarrow{\quad\xi\quad} & \bar{\mathbb{B}}(V,B)
\end{array}$$

is equivalent by σ_o to that of the exterior of diagram

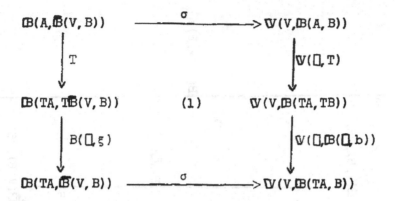

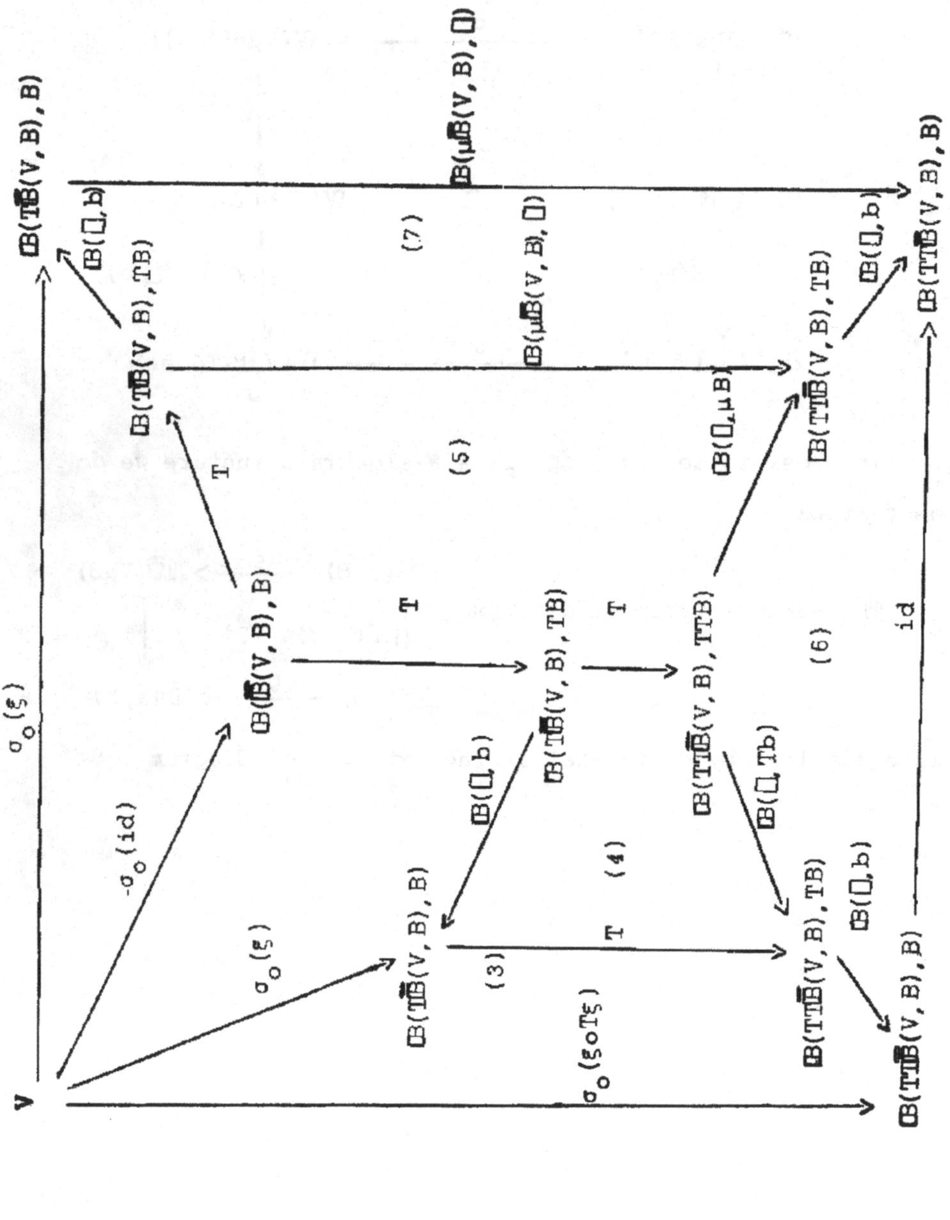
V
σo(ξ)
-σo(id)
σo(ξ)
σo(ξoTξ)
id
T
(3)
(4)
(5)
(6)
(7)

Diagram (4) commutes by V-functoriality of T, diagram (5) commutes by V-naturality of μ, diagram (6) because b is a $\mathbb{T}$-algebra structure and diagram (7) because $\mathbb{B}(-, -)$ is a bifunctor. Finally, diagram (1) at the level of sets with $A = T\bar{\mathbb{B}}(V,B)$ gives the commutativity of diagram (3). So diagram (2) commutes.

The commutativity of diagram

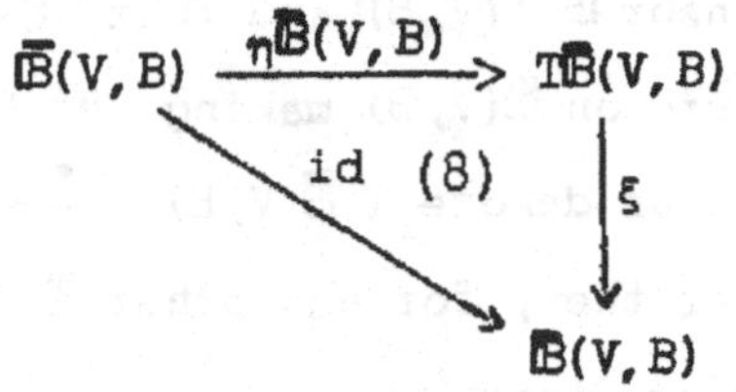

is equivalent by σ_o to that of the exterior of diagram.

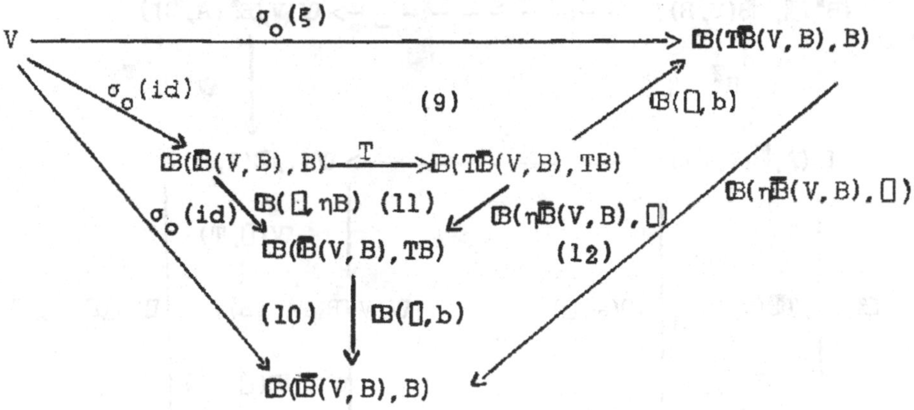

Diagram (9) commutes by definition of ξ, the commutativity of diagram (10) follows from the fact that b is a $\mathbb{T}$-algebra structure, diagram (11) commutes by V-naturality of η and diagram (12) because $\mathbb{B}(-,-)$ is a bifunctor. So diagram (8) commutes.

It remains now to prove that the $\mathbb{T}$-algebra defined above is a cotensor $\overline{\mathbb{B}^{\mathbb{T}}}(V,\bar{B})$ and that ξ is the unique $\mathbb{T}$-algebra structure on $\overline{\mathbb{B}}(V,B)$ making z be the equality.

Let us denote $T\,\overline{\mathbb{B}}(V,B) \xrightarrow{\xi} \overline{\mathbb{B}}(V,B)$ by $\overline{\overline{\mathbb{B}}(V,B)}$.
Consider then, for any other $\bar{A} \in \mathbb{B}^{\mathbb{T}}$, $\bar{A} = (TA \xrightarrow{a} A,)$ the following diagram:

$$
\begin{array}{ccccc}
\mathbb{B}^{\mathbb{T}}(\bar{A}, \overline{\overline{\mathbb{B}}(V,B)}) & & \overset{\bar{\sigma}}{\dashrightarrow} & & \mathbb{V}(V,\mathbb{B}^{\mathbb{T}}(\bar{A},\bar{B})) \\
\downarrow{\scriptstyle U^{\mathbb{T}}} & & & & \downarrow{\scriptstyle \mathbb{V}(\square,U^{\mathbb{T}})} \\
\mathbb{B}(A,\overline{\mathbb{B}}(V,B)) & & \xrightarrow{\quad\sigma\quad} & & \mathbb{V}(V,\mathbb{B}(A,B)) \\
\downarrow{\scriptstyle T} & & & & \downarrow{\scriptstyle \mathbb{V}(\square,T)} \\
\mathbb{B}(TA,T\overline{\mathbb{B}}(V,B)) & \Big\downarrow{\scriptstyle \mathbb{B}(a,\square)} & & \mathbb{V}(V,\mathbb{B}(TA,TB)) & \Big\downarrow{\scriptstyle \mathbb{V}(\square,\mathbb{B}(a,\square))} \\
\downarrow{\scriptstyle \mathbb{B}(\square,\xi)} & & & & \downarrow{\scriptstyle \mathbb{V}(\square,\mathbb{B}(\square,b))} \\
\mathbb{B}(TA,\overline{\mathbb{B}}(V,B)) & & \xrightarrow{\quad\sigma\quad} & & \mathbb{V}(V,\mathbb{B}(TA,B))
\end{array}
$$

Both columns are V-equalizers, the bottom square commutes by naturality of σ and the bottom hexagon is exactly diagram

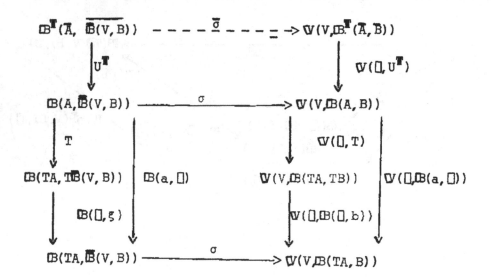

(1) and therefore commutes. So there is a unique isomorphism $\bar{\sigma}$ making the top square commutative. The easiest way to see that $\bar{\sigma}$ is a V-natural transformation in $\bar{A}$ is by letting (in the above diagram) $\bar{A}$ be (-), A be $U^{\mathbb{T}}(-)$ and a be

$$TU^{\mathbb{T}}(-) \xRightarrow{U^{\mathbb{T}}\epsilon} U^{\mathbb{T}}(-)$$ where ϵ is the counit $F^{\mathbb{T}}U^{\mathbb{T}} \xRightarrow{\epsilon} \mathrm{id}$.

All the arrows in the diagram became then double arrows and the columns pointwise V-equalizers of V-functors. Then $\bar{\sigma}$ is a V-natural transformation by definition. It can be seen, from the definition of z, ((1) page 24), that the commutativity of the upper square means exactly that z is the equality. Finally, any other $\mathbb{T}$-algebra structure on $\bar{\mathbb{B}}(U,B)$, ξ', which is a cotensor $\overline{\mathbb{B}^{\mathbb{T}}}(V,\bar{B})$ making z the equality, has to make the upper square (with its isomorphisms σ' in place of $\bar{\sigma}$) commutative. Note that now the hexagon (with ξ') does not necessarily commute, but the bottom square is still the same and so commutes. Then, putting this cotensor in place of $\bar{A}$, and using the diagram at the level of sets, the identity in the left upper corner goes down by the left column into ξ', while going around the perimeter, by the commutativity of the upper square, it goes into ξ (recall the definition of ξ).

Now, since the bottom square commutes, both ways are the same, and so $\xi = \xi'$. ∎

Let $\mathbb{B}$ any V-category and $\mathbb{T}$ a V-monad in $\mathbb{B}$. Then:

• Proposition II.4.8

a) Given a category $\mathbb{\Gamma}$ and a functor $\mathbb{\Gamma} \xrightarrow{\Gamma} \mathbb{B}^{\mathbb{T}}$. If $\underset{\overleftarrow{\lambda}}{\text{V-lim}}\, U^{\mathbb{T}}\Gamma_\lambda \xrightarrow{p_\lambda} U^{\mathbb{T}}\Gamma_\lambda$ exists, then $\underset{\overleftarrow{\lambda}}{\text{V-lim}}\, \Gamma_\lambda \xrightarrow{p_\lambda} \Gamma_\lambda$ also exists. Furthermore, there is a unique one such that:

$$\begin{array}{ccc} U^{\mathbb{T}}\, \underset{\overleftarrow{\lambda}}{\text{V-lim}}\, \Gamma_\lambda & = & \underset{\overleftarrow{\lambda}}{\text{V-lim}}\, U^{\mathbb{T}}\Gamma_\lambda \\ & {\searrow}^{U^{T}p_\lambda} \quad {\swarrow}_{p_\lambda} & \\ & U^{\mathbb{T}}\Gamma_\lambda & \end{array}$$

b) Given a V-category $\mathbb{C}$ and a V-bifunctor

$\mathbb{C}^{op} \otimes \mathbb{C} \xrightarrow{T} \mathbb{B}^{\mathbb{T}}$, if $\int_C U^{\mathbb{T}}\, T(C,C) \xrightarrow{pC} U^{\mathbb{T}}\, T(C,C)$ exists, then $\int_C T(C,C) \xrightarrow{pC} T(C,C)$ also exists. Furthermore, there is a unique one such that

$$\begin{array}{ccc} U^{\mathbb{T}}\int_C T(C,C) & = & \int_C U^{\mathbb{T}}\, T(C,C) \\ & {\searrow}^{U^{\mathbb{T}}pC} \quad {\swarrow}_{pC} & \\ & U^{\mathbb{T}}\, T(C,C) & \end{array}$$

$\mathbb{T}$-Algebras structures on $\underset{\overleftarrow{\lambda}}{\text{V-lim}}\, U^{\mathbb{T}}\Gamma_\lambda$ and on $\int_C U^{\mathbb{T}}\, T(C,C)$ are provided by the respective universal properties. The details of the proof follow similar lines to the ones developed in Propositions II.4.2 and II.4.3, and are left to the reader. ∎

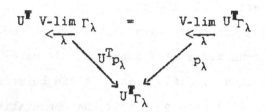

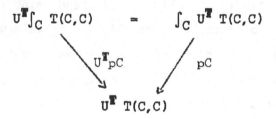

CHAPTER III

COMPLETE CATEGORIES

Section 1 V-continuous V-functors

Definition III.1.1

Let $\mathbb{A}$, $\mathbb{B}$ be V-categories and $\mathbb{B} \xrightarrow{G} \mathbb{A}$ be a V-functor; we say that G is V-continuous if it preserves V-limits, cotensors and ends. (see Definitions I.1.2, I.2.3 and I.3.3.)

Small Concepts

Given any V-category $\mathbb{A}$, a small category Γ and a small V-category $\mathbb{C}$, we will call the V-limit of a functor $\Gamma \longrightarrow \mathbb{A}$, if it exists, a small V-limit, and similarly, the end of a V-bifunctor $\mathbb{C}^{op} \otimes \mathbb{C} \longrightarrow \mathbb{A}$, if it exists, a small end.

Given any other V-category $\mathbb{B}$, we will call the V-limit of V-functors $\mathbb{B} \xrightarrow{\Gamma(\lambda,-)} \mathbb{A}$, $\Gamma \times \mathbb{B} \xrightarrow{\Gamma} \mathbb{A}$, a small V-limit of V-functors, and similarly, the end of V-functors $\mathbb{B} \xrightarrow{T(C,C,-)} \mathbb{A}$, $\mathbb{C} \otimes \mathbb{C}^{op} \otimes \mathbb{B} \xrightarrow{T} \mathbb{A}$, a small end of V-functors. Finally; given a V-functor $\mathbb{C} \xrightarrow{S} \mathbb{B}$, we will call the right Kan extension along S of any V-functor $\mathbb{C} \xrightarrow{T} A$, $\mathbb{B} \xrightarrow{Ran_S(T)} \mathbb{A}$ a small Kan extension.

It is clear that the criteria of existence for right Kan extensions (Theorems I.4.2 and I.4.3) relate small concepts to small concepts. A small Kan extension is pointwise if and only if it is a small pointwise end of V-functors.

Definition III.1.1.S

Let $\mathbb{A}$, $\mathbb{B}$ be V-categories and $\mathbb{B} \xrightarrow{G} \mathbb{A}$ be a V-functor, we say that G is small V-continuous if it preserves small V-limits, cotensors and small ends.

V-continuous V-functors are small V-continuous but not vice-versa. A small V-continuous V-functor sends V-monomorphisms into V-monomorphisms (since a morphism is a V-monomorphism if and only if its V-kernel pair consists of two identity maps). A composite of V-continuous (small V-continuous) V-functors is V-continuous (small V-continuous). A V-continuous (small V-continuous) V-functor does not in general preserve V-limits of V-functors (small V-limits of V-functors), ends of V-functors (small ends of V-functors) or right Kan extensions (small right Kan extensions), but it is clear that it does preserve the respective pointwise concepts.

Proposition III.1.1

For any V-category $\mathbb{A}$, the representable functors $\mathbb{A} \xrightarrow{\mathbb{A}(A,-)} \mathbb{V}$ are V-continuous.

Proof:

See remarks below definitions I.1.2 and I.3.3 and Proposition I.2.3. ∎

Proposition III.1.2

Given any two V-categories $\mathbb{A}$, $\mathbb{B}$ and a V-functor $\mathbb{B} \xrightarrow{G} \mathbb{A}$; G is V-continuous (small V-continuous) if and only if for every $A \in \mathbb{A}$, the V-functor $(\mathbb{B} \xrightarrow{G} \mathbb{A} \xrightarrow{\mathbb{A}(A,-)} \mathbb{V}) = (\mathbb{B} \xrightarrow{\mathbb{A}(A,G(-))} \mathbb{V})$ is V-continuous (small V-continuous).

Proof:

From Proposition II.1.1 the "only if" part follows trivially. The "if" part offers no difficulty. We will sketch it briefly and informally.

Let $\mathbb{\Gamma} \xrightarrow{\Gamma} \mathbb{B}$ be any functor and assume $V\text{-}\underset{\overleftarrow{\lambda}}{\lim}\, \Gamma_\lambda$ exists, then, for every $A \in \mathbb{A}$ $\mathbb{A}(A, G(V\text{-}\underset{\overleftarrow{\lambda}}{\lim}\, \Gamma_\lambda)) = \mathbb{A}(A, G(-))(V\text{-}\underset{\overleftarrow{\lambda}}{\lim}\, \Gamma_\lambda) \approx$

$$\approx V\text{-}\underset{\overleftarrow{\lambda}}{\lim}\, \mathbb{A}(A, G\Gamma_\lambda(-)),$$ and so,

just by definition of V-limits, we have $G(V\text{-}\underset{\overleftarrow{\lambda}}{\lim}\, \Gamma_\lambda) \approx V\text{-}\underset{\overleftarrow{\lambda}}{\lim}\, G\Gamma_\lambda$.

Let $V \in \mathbb{V}$ and $B \in \mathbb{B}$ and suppose $\bar{\mathbb{B}}(V,B)$ exists, then for every $A \in \mathbb{A}$

$\mathbb{A}(A, G\bar{\mathbb{B}}(V, B)) = \mathbb{A}(A, G(-))(\bar{\mathbb{B}}(V,B)) \approx \mathbb{V}(V, \mathbb{A}(A, GB))$, hence, just by definition of cotensors we have $G\bar{\mathbb{B}}(V,B) \approx \bar{\mathbb{A}}(V, GB)$.

Ends can be checked in a similar way. ∎

Proposition III.1.3

For any two V-categories $\mathbb{A}$, $\mathbb{B}$, and a V-functor $\mathbb{B} \xrightarrow{G} \mathbb{A}$; if G has a V-left adjoint F, then G is V-continuous.

Proof:

For every $A \in \mathbb{A}$, $\mathbb{A}(A, G(-)) \cong \mathbb{B}(FA, -)$ (V-natural), then from Propósition III.1.1 and Proposition III.1.2 we are done.

∎

Having a V-left adjoint implies and even stronger continuity property.

Proposition III.1.4

For any two V-categories $\mathbb{A}$, $\mathbb{B}$ and a V-functor $\mathbb{B} \xrightarrow{G} \mathbb{A}$; if G has a V-left adjoint F, then G preserves any V-limit, end or cotensor of V-functors with codomain $\mathbb{B}$.

Proof:

As in the case of right Kan extensions (Proposition I.4.2) it readily follows from Psoposition 0.1. ∎

Proposition III.1.5

Let $V \in \mathbb{V}$ be an object of $\mathbb{V}$, then, for any V-category $\mathbb{A}$, if it exists, the V-functor $\mathbb{A} \xrightarrow{\bar{\mathbb{A}}(V,-)} \mathbb{A}$ is V-continuous.

Proof:

For every $A \in \mathbb{A}$,

$\mathbb{A}(A, \bar{\mathbb{A}}(V,-)) \cong \mathbb{V}(V, \mathbb{A}(A,-)) = \mathbb{V}(V,-) \circ \mathbb{A}(A,-)$ (V-natural).

Then by Proposition III.1.1 and Proposition III.1.2 we are done.

∎

Proposition III.1.6

Pointwise V-limits (small pointwise V-limits), pointwise ends (small pointwise ends) and pointwise cotensors of V-continuous (small V-continuous) V-functors are V-continuous (small V-continuous).

The proof splits in several parts, that, in order to avoid confusion, we state as three different lemmas. We develop only the unrestricted case, but it is clear that for the small concepts exactly the same proof applies.

Lemma 1 (V-limits)

We refer to the situation (or data) of Definition I.1.3 (page 11). Then:
If for every $\lambda \in \mathbb{\Gamma}$, $\mathbb{B} \xrightarrow{\Gamma(\lambda,-)} \mathbb{A}$ is V-continuous, then so is $\mathbb{B} \xrightarrow{V\text{-}\varprojlim_{\lambda} \Gamma(\lambda,-)} \mathbb{A}$. (when it is pointwise.)

Proof:

a) Preservation of V-limits.

Let $\mathbb{X} \xrightarrow{X} \mathbb{B}$ be any functor and suppose the V-limit of X exists. Consider $\mathbb{\Gamma} \times \mathbb{X} \xrightarrow{id \times X} \mathbb{\Gamma} \times \mathbb{B} \xrightarrow{\Gamma} \mathbb{A}$, then use formula (1) (page 10).

b) Preservation of cotensors.

Let $V \in \mathbb{V}$ and $B \in \mathbb{B}$ and suppose $\bar{\mathbb{B}}(V, B)$ exists. Then:

$$\underset{\overleftarrow{\lambda}}{\text{V-lim}}\, \Gamma(\lambda, \bar{\mathbb{B}}(V,B)) \xrightarrow[\overleftarrow{\lambda}]{\text{V-lim } z_\lambda} \underset{\overleftarrow{\lambda'}}{\text{V-lim}}\, \bar{\mathbb{A}}(V,\Gamma(\lambda,B)) \approx$$

(by Proposition III.1.5) $\approx \bar{\mathbb{A}}\,(V, \underset{\overleftarrow{\lambda}}{\text{V-lim}}\, \Gamma(\lambda, B))$.

c) Preservation of ends.

Let $\mathbb{C}^{op} \otimes \mathbb{C} \xrightarrow{T} \mathbb{B}$ be any V-bifunctor and suppose the end of T exists.

Consider $\mathbb{C}^{op} \otimes \mathbb{C} \times \Gamma \xrightarrow{T \times id} \mathbb{B} \times \Gamma \xrightarrow{\Gamma} \mathbb{A}$ then use formula (1) (page 38). ∎

<u>Lemma 2</u> (cotensors)

We refer to the situation (or data) of Definition I.2.4 (page 27). Then:

If G is V-continuous, then so is $\bar{\mathbb{A}}^{\mathbb{B}}(V, G)$ (when it is pointwise).

<u>Proof</u>:

For every $A \in \mathbb{A}$

$$\mathbb{A}(A, \bar{\mathbb{A}}^{\mathbb{B}}(V, G)(-)) = \mathbb{A}(A, \bar{\mathbb{A}}(V, G(-))) \cong \mathbb{V}(V, \mathbb{A}(A, G(-))) =$$

$= \mathbb{V}(V, -) \circ \mathbb{A}(A, -) \circ G$. (V-natural), then by Proposition III.1.1 and Proposition III.1.2 we are done. ∎

Lemma 3 (ends).

We refer to the situation (or data) of Definition I.3.4 (page 33) Then:

If for every $C \in \mathbb{C}$, $\mathbb{B} \xrightarrow{T(C,C,-)} \mathbb{A}$ is V-continuous, then so is $\mathbb{B} \xrightarrow{\int_C T(C,C,-)} \mathbb{A}$. (when it is pointwise).

Proof:

a) Preservation of V-limits.

Let $\mathbb{\Gamma} \xrightarrow{\Gamma} \mathbb{B}$ be any functor and suppose the V-limit of Γ exists. Consider:

$$\mathbb{C}^{op} \otimes \mathbb{C} \times \mathbb{\Gamma} \xrightarrow{id \otimes id \times \Gamma} \mathbb{C}^{op} \otimes \mathbb{C} \times \mathbb{B} \longrightarrow \mathbb{C}^{op} \otimes \mathbb{C} \otimes \mathbb{B} \xrightarrow{T} \mathbb{A}$$

then use formula (1) (page 38).

b) Preservation of cotensors.

Let $V \in \mathbb{V}$ and $B \in \mathbb{B}$ and suppose $\bar{\bar{\mathbb{B}}}(V,B)$ exists. Then

$$\int_C T(C,C,\bar{\bar{\mathbb{B}}}(V,B)) \xrightarrow{\int_C zC} \int_C \bar{\bar{\mathbb{A}}}(V,T(C,C,B)) \approx \bar{\bar{\mathbb{A}}}(V, \int_C T(C,C,B)),$$

the latter isomorphism by Proposition III.1.5 .

c) Preservations of ends.

Let $\mathbb{D}^{op} \otimes \mathbb{D} \xrightarrow{S} \mathbb{B}$ be any V-bifunctor and suppose the end of S exists. Consider:

$$\mathbb{C}^{op} \otimes \mathbb{C} \otimes \mathbb{D}^{op} \otimes \mathbb{D} \xrightarrow{id \otimes id \otimes S} \mathbb{C}^{op} \otimes \mathbb{C} \otimes \mathbb{B} \xrightarrow{T} \mathbb{A}.$$

then use formula (1) (page 36). ∎

<u>Proposition III.1.7</u>

Given a V-full-and-faithful functor $\mathbb{C} \xrightarrow{R} \mathbb{A}$, $\mathbb{C}$, $\mathbb{A}$ any V-categories, let $\mathbb{B}$ be the V-full subcategory whose objects are those objects A of $\mathbb{A}$ for which the V-functor $\mathbb{A}^{op} \xrightarrow{\mathbb{A}(R(-),\, \mathbb{A})} \mathbb{V}$ is V-continuous. Then:

a) R factors through $\mathbb{B}$,

$$\begin{array}{ccc} \mathbb{C} & \xrightarrow{R} & \mathbb{A} \\ & {\scriptstyle S}\searrow \quad \nearrow & \\ & \mathbb{B} & \end{array}$$

(i.e., for every $C \in \mathbb{C}$, RC belongs to $\mathbb{B}$) and the V-full-and-faithful V-functor $\mathbb{C} \xrightarrow{S} \mathbb{B}$ is V-cocontinuous.

b) $\mathbb{B}$ is closed under all V-limits, cotensors and ends which might exist in $\mathbb{A}$.

Proof:

a) Let $C \in \mathbb{C}$, then $\mathbb{A}(R(-), RC) \approx \mathbb{C}(-, C) = \mathbb{C}^{op}(C, -)$, so by Proposition III.1.1 SC belongs to $\mathbb{B}$. The dual of Proposition III.1.2 gives all that remains to be proven.

b) It readily follows immediately from Proposition III.1.6 and Proposition III.1.1. ∎

Note that $\mathbb{B}$ closed under V-limits, cotensors and ends means that $\mathbb{B}$ has at least as many V-limits, cotensors or ends as $\mathbb{A}$ has, but it does not mean that it cannot have more, and so we cannot conclude that the inclusion $\mathbb{B} \longrightarrow \mathbb{A}$ is V-continuous. Note also that we can define a (possibly larger) V-full subcategory $\mathbb{B}'$ in a similar way but requiring for objects $A \in \mathbb{B}'$ that the V-functor $\mathbb{A}^{op} \xrightarrow{\mathbb{A}(R(-),A)} \mathbb{V}$ be only small V-continuous. In this case, the V-functor $\mathbb{C} \xrightarrow{S'} \mathbb{B}'$ is only small V-cocontinuous and $\mathbb{B}'$ is closed under small V-limits, cotensors and small ends. We will not use, however, this version of Proposition III.1.7. Note also that, since a pointwise V-subfunctor of a V-continuous (small V-continuous) V-functor need not be V-continuous (small V-continuous) it follows (from the way in which Proposition III.1.7. is proved) that in no case $\mathbb{B}$ is closed under V-sub-objects.

<u>Proposition III.1.8</u>

Given a V-codense V-full-and-faithful V-functor $\mathbb{C} \xrightarrow{R} \mathbb{A}$, $\mathbb{C}$, $\mathbb{A}$ any V-categories, $\mathbb{A}$ cotensored, then R is necessarily V-cocontinuous.

<u>Proof</u>:

For every $A \in \mathbb{A}$, $A \cong \mathrm{Ran}_R(R)(A) = \int_C \overline{\mathbb{A}}\,(\mathbb{A}(A,RC),RC)$ (and the end exists). Let $\mathbb{B}$ be as in the previous proposition. Then, $RC \in \mathbb{B}$ for every $C \in \mathbb{C}$ and therefore $A \in \mathbb{B}$. So $\mathbb{A} = \mathbb{B}$. ∎

It is clear that there are concepts dual to the ones developed in this section, with the respective dual statements.

Section 2 V-complete V-categories

Definition III.2.1

A V-category $\mathbb{A}$ is said to be V-complete if it has all small V-limits and is cotensored. In other words, if for every $A \in \mathbb{A}$ and $V \in \mathbb{V}$ the cotensor $\overline{\mathbb{A}}(V,A)$ exists, if for every functor $\mathbb{\Gamma} \xrightarrow{\Gamma} \mathbb{A}$ with $\mathbb{\Gamma}$ small the limit of Γ exists and if for every $A \in \mathbb{A}$ the functor $\mathbb{A}(A,-)$ preserves small limits.

It is clear that V-complete V-categories should have small ends. This is actually the case:

Proposition III.2.1

Given a V-bifunctor $\mathbb{C}^{op} \otimes \mathbb{C} \xrightarrow{T} \mathbb{A}$, where $\mathbb{C}$ is small and $\mathbb{A}$ V-complete, then the end of T exists, and it is a small V-limit of cotensors.

Proof:

Just observe that the diagram involved in Proposition I.3.5 (page 39) is small when $\mathbb{C}$ is so. ∎

Corollary

A V-functor $\mathbb{B} \xrightarrow{G} \mathbb{A}$, between V-complete V-categories is small V-continuous if and only if it preserves small V-limits and cotensors. ∎

It follows from definition III.2.1 that in V-complete V-categories every monomorphism is a V-monomorphism. Also, V-natural transformations of V-functors with V-complete codomain are monomorphism if and only if are pointwise V-monomorphisms. Cotensors, small V-limits and small ends of V-functors with V-complete codomain exists and are pointwise. From Theorem I.4.2 it follows that this is also true for small right Kan extensions. Explicitly:

<u>Theorem II.2.1</u> (Kan)

Let $\mathbb{C} \xrightarrow{S} \mathbb{B}$ be a V-functor where $\mathbb{C}$ is a small V-category. If $\mathbb{A}$ is a V-complete V-category, then for any V-functor $\mathbb{C} \xrightarrow{T} \mathbb{A}$, the right Kan extension of T along S, $\mathbb{B} \xrightarrow{\mathrm{Ran}_S(T)} \mathbb{A}$ exists and is pointwise. ∎

Any V-functor $\mathbb{C} \xrightarrow{S} \mathbb{B}$ from a small V-category $\mathbb{C}$ into a V-complete V-category $\mathbb{B}$ is tractable. It is V-dense if and only if $id_{\mathbb{B}} \xrightarrow{\eta} \mathbf{T}_S$ is an isomorphis and it is V-cogenerating if and only if $id_{\mathbb{B}} \xrightarrow{\eta} \mathbf{T}_S$ is a monomorphism.

Dually than in Definition III.2.1, we define a V-category $\mathbb{A}$ to be V-cocomplete if it has all small V-colimits and is tensored.

An immediate consequence of these definitions is that the base category $\mathbb{V}$, being tensored and cotensored, is V-complete or V-cocomplete if and only if it is complete or cocomplete as a set-based category.

Our next two propositions concern existence and preservation by small-continuous V-functors of large right Kan extensions with V-complete codomain. They are preliminary steps necessary for our proof of the V-version of the special adjoint functor theorem <u>within</u> the V-context.

A V-category is said to be V-well powered if the class of V-sub-objects of any object is a set. V-complete V-categories are V-well powered if and only if they are well powered in the ordinary sense.

The property of admitting a right Kan extension is hereditary for V-functors with a V-well powered V-complete codomain. Explcitly:

Let $\mathbb{C} \xrightarrow{S} \mathbb{B}$ any V-functor, $\mathbb{C}$, $\mathbb{B}$ any V-categories

<u>Proposition III.2.2</u>

Given a V-well powered V-complete V-category $\mathbb{A}$, V-functors $\mathbb{C} \xrightarrow{T'} \mathbb{A}$, $\mathbb{C} \xrightarrow{T} \mathbb{A}$ and a monomorphic V-natural transformation $T \overset{\varphi}{\Longrightarrow} T'$. Then: If $\mathbb{B} \xrightarrow{Ran_S(T')} \mathbb{A}$ exists, then so does $\mathbb{B} \xrightarrow{Ran_S(T)} \mathbb{A}$.

We have the configuration:

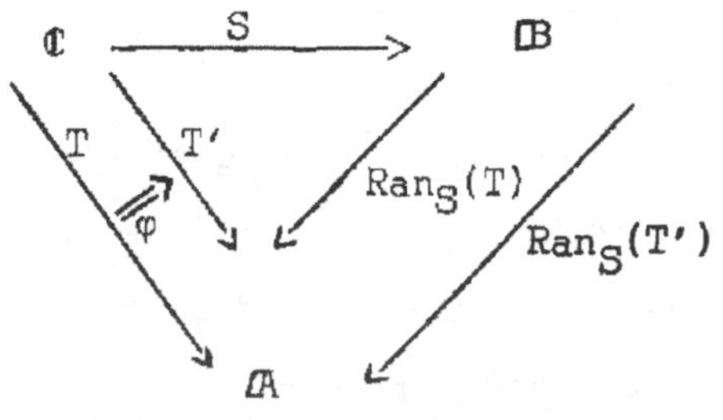

Proof:

By Theorem I.4.3 all we have to do is prove that the end of the V-functors $\mathbb{B} \xrightarrow{\bar{\mathbb{A}}(\mathbb{B}(-,SC),\ TC)} \mathbb{A}$ exists. Consider the diagram:

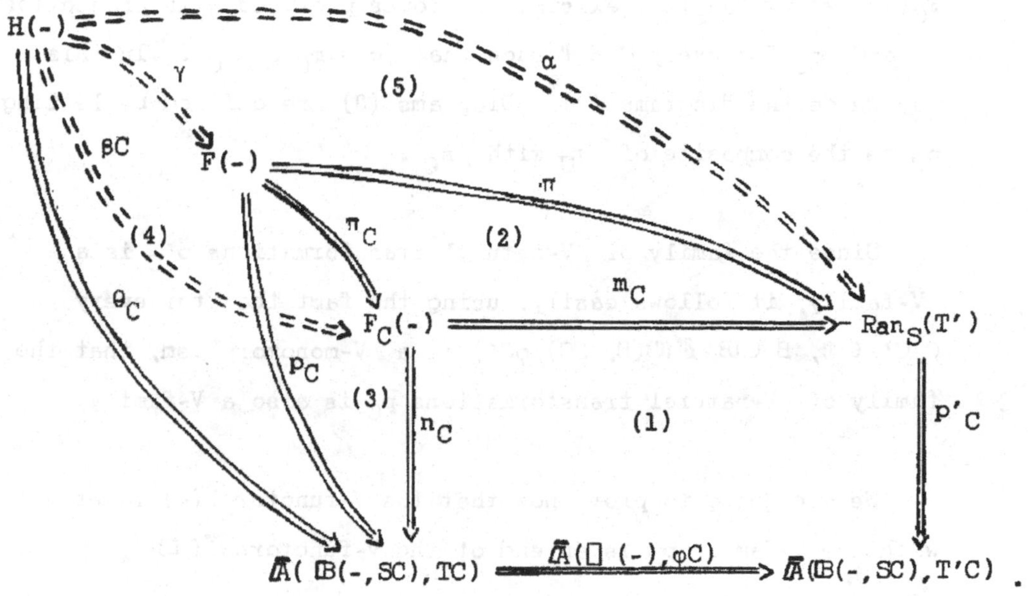

We define diagrams (1) as a V-meet of V-functors. These meets exist and are pointwise because $\mathbb{A}$ is V-complete. Since φ_C is a V-monomorphism for every C; it follows (from Proposition III.1.5) that $\bar{\mathbb{A}}(\square(B), \varphi_C)$ is a V-monomorphism for every $B \in \mathbb{B}$, $C \in \mathbb{C}$. So by Proposition I.1.2 $m_C B$ is a V-monomorphism for every $B \in \mathbb{B}$, $C \in \mathbb{C}$. Then; with a fixed B, (since $\mathbb{A}$ is V-well powered) the intersection of all the V-sub-objects

$F_C(B) \xrightarrow{m_C B} Ran_S(T')(B)$ exists in $\mathbb{A}$ (i.e. the large V-limit over the class of objects of $\mathbb{C}$ of the diagram $F_C(B) \xrightarrow{m_C B} Ran_S(T')(B)$ exists). Since this happens for every $B \in \mathbb{B}$; the V-limit of V-functors of the diagram $F_C(-) \overset{m_C}{\Longrightarrow} Ran_S(T')$ exists. It comes provided with projections π and π_C for every $C \in \mathbb{C}$ such that $\pi = m_C \circ \pi_C$. In this way we define diagrams (2). Diagrams (3) are defined by letting p_C be the composite of π_C with n_C.

Since the family of V-natural transformations p'_C is a V-family, it follows easily, using the fact that for every $C, C' \in \mathbb{C}$, $B \in \mathbb{B}$ $\bar{\mathbb{A}}(\mathbb{B}(B, SC), \varphi C')$ is a V-monomorphism, that the family of V-natural transformations p_C is also a V-family.

We are going to prove now that the V-functor $F(-)$ together with the V-family p_C is an end of the V-functors $\bar{\mathbb{A}}(\mathbb{B}(-,SC),TC)$.

Let $H \overset{\theta_C}{\Longrightarrow} \bar{\mathbb{A}}(\mathbb{B}(-,SC),TC)$ be any other V-family of V-natural transformations. Then, composing with $\bar{\mathbb{A}}(\square\,(-), \varphi C)$ we obtain a V-family into the $\bar{\mathbb{A}}(\mathbb{B}(-,SC),T'C)$'s. Then, because $Ran_S(T')$ is the end of those V-functors, a V-natural transformation α appears, unique making the outer perimeter commutative. Then, since $F_C(-)$ is a meet of V-functors, we produce β_C for every $C \in \mathbb{C}$, unique such that $n_C \circ \beta_C = \theta_C$ and $m_C \circ b_C = \alpha$.

Finally, by definition of F(-); there is a unique V-natural transformation $H(-) \xRightarrow{\gamma} F(-)$ making diagrams (4) and (5) commutative. Any other $H \xRightarrow{\gamma'} F$ making diagrams (4) commutative, by the uniqueness of α also makes diagram (5) commutative. Hence it has to be equal to γ. ∎

Note that neither $Ran_S(T')$ or $Ran_S(T)$ in the above proposition need to be pointwise. However, for every $C \in \mathbb{C}$ the V-meet of V-functors (diagrams (1)) is pointwise and the large V-limit of V-functors (intersection) (diagram (2)) can be computed pointwise by means of small (perhaps over different diagrams for each $B \in \mathbb{B}$) V-limits in $\mathbb{A}$. Having this in mind, and recalling that any small V-continuous V-functor sends V-monomorphisms into V-monomorphism, it is not difficult to prove the following:

Proposition III.2.3

In the situation of Proposition III.2.2, any small V-continuous V-functor $\mathbb{A} \xrightarrow{G} \mathbb{A}'$ which preserves $Ran_S(T')$ also preserves $Ran_S(T)$. ∎

Note that there are no completeness or well-power requirements for $\mathbb{A}'$.

Theorem III.2.2 (Special Adjoint Functor Theorem)

Let $\mathbb{B}$ be a V-well powered V-complete V-category admitting a small V-cogenerating V-functor $\mathbb{C} \xrightarrow{S} \mathbb{B}$ (in particular, if

$\mathbb{B}$ has a V-cogenerator). Then, every small V-continuous V-functor $\mathbb{B} \xrightarrow{G} \mathbb{A}$ into any other V-category $\mathbb{A}$ has a V-left adjoint.

<u>Proof</u>:

We have the following configuration:

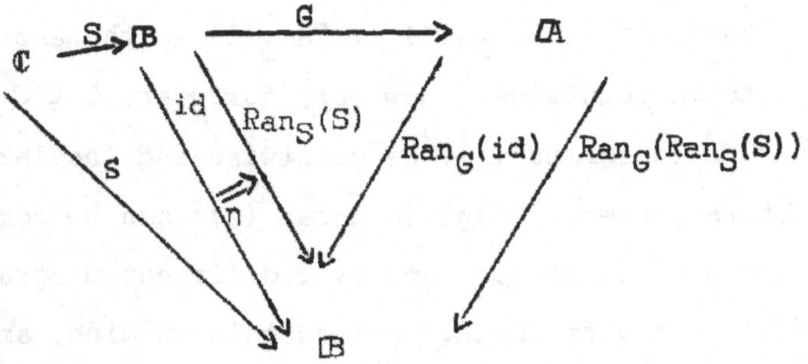

Since $\mathbb{C}$ is small, $Ran_S(S)$ and $Ran_{GS}(S)$ exist and are pointwise. The small V-continuity of G implies that G preserves them. Then by Proposition I.4.1 $Ran_G(Ran_S(S))$ exists and it is preserved by G. Since η is a monomorphism we use Propositions III.2.2 and III.2.3 to deduce that $Ran_G(id)$ exists and is preserved by G. Then, Theorem I.4.1 completes the proof. ∎

Since in the general V case a V-cogenerator is far away from being a real cogenerator, the proof of the above Theorem suggests that the special Adjoint Functor Theorem is due to the pure formal properties of the concepts involved rather that to the substantial fact that generators provide solution sets. However, as opposed to the more formal adjoint functor theorem recorded below, the substantial requirement that $\mathbb{B}$ be well-powered (or some strong completeness requirement in $\mathbb{B}$) retains its essential role, as it is clearly seen in the proof of Proposition III.2.2.

Theorem III.2.3

Let $\mathbb{B}$ a V-complete V-category admitting a small V-codense V-functor $\mathbb{C} \xrightarrow{S} \mathbb{B}$. Then, every small V-continuous V-functor $\mathbb{B} \xrightarrow{G} \mathbb{A}$ into any other V-category $\mathbb{A}$ has a V-left adjoint $\mathbb{A} \xrightarrow{F} \mathbb{B}$. Moreover; F can be computed by means of the formula:

$$(1) \quad FA = \int_C \overline{\mathbb{B}}(\mathbb{A}(A,GSC),SC) = \mathrm{Ran}_{GS}(S)$$

Proof:

Following the same lines as in Theorem III.2.2, notice that in this case $\mathrm{Ran}_S(S) = \mathrm{id}$ and so Propositions III.2.2 and III.2.3 are not needed. Proposition I.4.1 and Theorem I.4.1 give the formula. ∎

When the base category $\mathbb{V}$ is complete (hence V-complete), V-categories with enough compactness properties are V-cocomplete. In particular, the previous two theorems guarantee the following two results.

•• Proposition III.2.4

Let $\mathbb{B}$ be a V-complete V-category admitting a small V-codense V-functor $\mathbb{C} \xrightarrow{S} \mathbb{B}$. Then $\mathbb{B}$ is V-cocomplete.

•• Proposition III.2.5

Let $\mathbb{B}$ be a V-well-powered V-complete V-category admitting a small V-cogenerating V-functor $\mathbb{C} \xrightarrow{S} \mathbb{B}$. Then $\mathbb{B}$ is V-cocomplete.

Proof:

Let $B \in \mathbb{B}$ any object of $\mathbb{B}$. $\mathbb{B} \xrightarrow{\mathbb{B}(B,-)} \mathbb{V}$ is V-continuous, hence it has a V-left adjoint. So $\mathbb{B}$ is tensored.

Let $\mathbb{\Gamma} \xrightarrow{\Gamma} \mathbb{B}$ be any functor into $\mathbb{B}$, where $\mathbb{\Gamma}$ is any small category. Consider the V-functors $\mathbb{B} \xrightarrow{\mathbb{B}(\Gamma_\lambda,-)} \mathbb{V}$. Since $\mathbb{V}$ is V-complete, the V-limit of V-functors exists (pointwise) and from Proposition III.1.6 it follows that it is V-continuous. Hence it has a V-left adjoint $\mathbb{V} \xrightarrow{F} \mathbb{B}$. Then:

$$\begin{array}{ccc} \mathbb{B}(F(I),-) & \cong & \underset{\overleftarrow{\lambda}}{\text{V-lim}}\, \mathbb{B}(\Gamma_\lambda,-) \\ & {\scriptstyle \mathbb{B}(i_\lambda,-)} \searrow \quad \swarrow {\scriptstyle p_\lambda} & \\ & \mathbb{B}(\Gamma_\lambda,-) & \end{array} ,$$

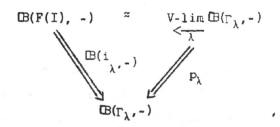

where $\Gamma_\lambda \xrightarrow{i_\lambda} F(I)$ is provided by the representation theorem. Since the V-limit is pointwise, $\Gamma_\lambda \xrightarrow{i_\lambda} F(I)$ is a V-colimit of Γ exactly by definition of V-colimits. ∎

Given any V-monad $\mathbb{T}$ in a V-complete V-category $\mathbb{A}$, it follows from Propositions II.4.7 and II.4.8 that the V-category of algebras over $\mathbb{T}$ is also V-complete. More generally:

• Proposition III.2.6

Given any monadic V-functor $\mathbb{B} \xrightarrow{G} \mathbb{A}$, if $\mathbb{A}$ is V-complete, then so is $\mathbb{B}$. (For the definition of monadic see Theorem II.2.1) ∎

Two expected properties of V-reflexives V-full sub-categories are that they are V-complete and V-cocomplete when the larger V-category is so.

Let $\mathbb{B} \xrightarrow{G} \mathbb{A}$ a V-full-and-faithful V-functor having a V-left adjoint, then:

Proposition III.2.7

If $\mathbb{A}$ is V-cocomplete, then so is $\mathbb{B}$.

Proposition III.2.8

If $\mathbb{A}$ is V-complete, then so is $\mathbb{B}$.

The proof of Proposition III.2.7 offers no difficulty; for V-colimits we do as in ordinary set-based category theory but using the representable V-functors into $\mathbb{V}$. For tensors; given $V \in \mathbb{V}$ and $B \in \mathbb{B}$, we define $V \otimes_{\mathbb{B}} B = F(V \otimes_{\mathbb{A}} GB)$, where F is a V-left-adjoint to G. The following chain of isomorphisms gives the result.

$$\mathbb{B}(F(V \otimes_{\mathbb{A}} GB), B') \approx \mathbb{A}(V \otimes_{\mathbb{A}} GB, GB') \approx \mathbb{V}(V, A(GB, GB') \approx$$
$$\approx \mathbb{V}(V, \mathbb{B}(B, B')) \quad .$$

Proposition III.2.8 is not so easy to prove directly, but it follows from Proposition III.2.6 and II.2.1. ∎

<u>Section 3</u> <u>Relative V-completions</u>

We consider (in this paper) the problem of completions of small V-categories to be the following:

Whether, given a small V-category $\mathbb{C}$, there is a V-complete and V-cocomplete V-category $\mathbb{B}$ and a V-full-and-faithful V-functor $\mathbb{C} \xrightarrow{S} \mathbb{B}$ such that, those V-limits and V-colimits, end and coends, and tensors and cotensors which exist in $\mathbb{C}$, remain the same when taken in $\mathbb{B}$; that is, such that $\mathbb{C} \xrightarrow{S} \mathbb{B}$ is V-continuous and V-cocontinuous. Moreover, we require also that $\mathbb{B}$ be "small enough" in relation to $\mathbb{C}$ in the sense that $\mathbb{C} \xrightarrow{S} \mathbb{B}$ should be V-codense and V-generating, or, dually, V-dense and V-cogenerating. We will call such a $\mathbb{B}$ (that is, such a $\mathbb{C} \xrightarrow{S} \mathbb{B}$) a <u>V-completion</u> of $\mathbb{C}$.

The problem has an affirmative answer <u>when the base category $\mathbb{V}$ is complete and well powered</u> (hence V-complete and V-well powered).

In this section we develop two different techniques by means of which we obtain such a V-completion. In doing so, we start by <u>assuming</u> that there is a V-functor $\mathbb{C} \xrightarrow{R} \mathbb{A}$ into a V-category $\mathbb{A}$ with certain properties and we then obtain a factorization

$$\begin{array}{ccc} \mathbb{C} & \xrightarrow{R} & \mathbb{A} \\ & {\scriptstyle S}\searrow \quad \nearrow & \\ & \mathbb{B} & \end{array}$$

where $\mathbb{C} \xrightarrow{S} \mathbb{B}$ is a V-completion. This V-completion <u>depends</u> on the starting data $\mathbb{C} \xrightarrow{R} \mathbb{A}$, and because of this we call it a <u>relative V-completion</u> of $\mathbb{C}$ with respect to R. Since we start by assuming the existence of a V-category $\mathbb{A}$ and a V-functor $\mathbb{C} \xrightarrow{R} \mathbb{A}$ satisfying certain properties, this relative V-completions don't prove the existence of a V-completion for $\mathbb{C}$ unless we can show the existence of a $\mathbb{C} \xrightarrow{R} \mathbb{A}$ first. To do this it is necessary to develop the construction of V-functor categories, which we do in the next chapter, where we state the (absolute) theorems of existence of V-completions as immediate corollaries of the results of this section.

First Relative V-Completion

Proposition III.3.1

Let $\mathbb{A}$ be a V-complete V-category and $\mathbb{C} \xrightarrow{R} \mathbb{A}$ any tractable V-full-and-faithful V-functor. Let $\mathbb{B}$ be the intersection of the V-full sub-categories defined in Propositions II.3.2 and III.1.7. That is, $\mathbb{B}$ is the V-full subcategory whose objects are those objects $A \in \mathbb{A}$ for which $A \xrightarrow{\eta A} T_R(A)$ is a V-monomorphism and the V-functor $\mathbb{C}^{op} \xrightarrow{\mathbb{A}(R(-),A)} \mathbb{V}$ is V-continuous. Then:

a) R factors through $\mathbb{B}$,

$$\begin{array}{ccc} \mathbb{C} & \xrightarrow{R} & \mathbb{A} \\ & {\scriptstyle S}\searrow \quad \nearrow & \\ & \mathbb{B} & \end{array}$$

and the V-full-and-faithful V-functor $\mathbb{C} \xrightarrow{S} \mathbb{B}$ is V-cocontinuous and V-cogenerating.

b) $\mathbb{B}$ is V-complete and the V-inclusion $\mathbb{B} \longrightarrow \mathbb{A}$ is small V-continuous.

Proof:

Write $\mathbb{B} = \mathbb{B}_o \cap \mathbb{B}_1$ where $\mathbb{B}_o$ is as in Proposition II.2.2 and $\mathbb{B}_1$ as in Proposition III.1.7. It is clear that R factors through $\mathbb{B}$ and that $\mathbb{C} \xrightarrow{S} \mathbb{B}$ is V-cocontin ous (see proof of III.1.7).

Consider the end $Ran_R(R)(B) = \int_C \overline{\mathbb{A}}(\mathbb{A}(B,RC), RC)$ (for any $B \in \mathbb{B}$).

For every $C \in \mathbb{C}$, $\bar{\mathbb{A}}(\mathbb{A}(B,RC), RC)$ belongs to $\mathbb{B}$ and $\mathbb{B}(\mathbb{A}(B,RC), RC) = \bar{\mathbb{A}}(\mathbb{A}(B,RC), RC)$. On the other hand, as in Proposition II.2.2, since $\eta\mathrm{Ran}_R(R)(B)$ splits, the end belongs to $\mathbb{B}_o$, and, since $\mathbb{B}_1$ is closed under all (large) ends, it also belongs to $\mathbb{B}_1$. So, $\mathrm{Ran}_S(S)$ exists (pointwise) and for every $B \in \mathbb{B}$, $\mathrm{Ran}_S(S)(B) = \mathrm{Ran}_R(R)(B)$. Hence, S is V-cogenerating. Finally, since $\mathbb{A}$ is V-complete, any small V-limit (end) is conserved by $\mathrm{Ran}_R(R)$, hence, (besides being closed under cotensors) $\mathbb{B}_o$ is closed under small V-limits (ends). Then so is $\mathbb{B}$. Hence, since $\mathbb{A}$ is V-complete, $\mathbb{B}$ is also V complete and the V-inclusion $\mathbb{B} \longrightarrow \mathbb{A}$ is small V-continuous. ∎

Remark IV.3.1

If $\mathbb{C}$ is small, $\mathbb{B}$ as in the above proposition is exactly the V-full subcategory whose objects are all V-subobjects A of small V-limits of cotensors of objects of the form RC, $C \in \mathbb{C}$ and for which the V-functor $\mathbb{C}^{op} \xrightarrow{\mathbb{A}(R(-),\, A)} \mathbb{V}$ is V-continuous.

Proof:

Just observe that for every $B \in \mathbb{B}$, the end $\mathrm{Ran}_R(R)(B)$ is small. Then the result follows from the considerations made above Remark II.3.1. ∎

Theorem III.3.1 a) and •• Theorem III.3.1 b) ($1^{\underline{o}}$ Relative Completion)

Let $\mathbb{C}$ be any small V-category and $\mathbb{C} \xrightarrow{R} \mathbb{A}$ a V-full-and-faithful V-dense (hence V-continuous) V-functor into a

V-complete V-well powered V-category. (For a) we assume $\mathbb{A}$ to be also V-cocomplete, for b) we assume the base category $\mathbb{V}$ to be complete). Then, $\mathbb{C} \xrightarrow{S} \mathbb{B}$ (as in the previous Proposition) is a V-full-and-faithful V-continuous and V-cocontinuous V-cogenerating and V-dense V-functor into a V-complete and V-cocomplete V-category.

<u>Proof</u>:

That S is V-full-and-faithful, V-cocontinuous and V-cogenerating and that $\mathbb{B}$ is V-complete has been established in the previous Proposition. Recalling statement b) of the same Proposition, that $\mathbb{B}$ is also V-cocomplete follows then from Theorem III.2.2 and Proposition III.2.7 in the a) case, and from Proposition III.2.5 in the b) case. Finally, the following chain of V-natural isomorphisms shows that S is V-dense, and so, in view of Proposition III.1.8, also V-continuous.

Write $\mathbb{A} \xrightarrow{F} \mathbb{B}$ for the V-left adjoint of the V-inclusion $\mathbb{B} \xrightarrow{I} \mathbb{A}$ then:

$$id \approx FI \approx F\,Lan_R(R)I \approx Lan_R(FR)I \approx Lan_R(S)I \approx Lan_{IS}(S)I \approx$$

$$\approx Lan_I(Lan_S(S))I \approx Lan_S(S)$$

(Notice that $FR = FIS \stackrel{\approx}{=} S$) ∎

Second Relative V-completion

•• Proposition III.3.2

Given a V-functor $\mathbb{C} \xrightarrow{R} \mathbb{A}$ from a small V-category into a V-complete V-category, it gives rise to the following tower of V-categories and V-functors:

a) For every ordinal γ there is a V-category $\mathbb{A}_\gamma$ and a V-functor $\mathbb{C} \xrightarrow{S_\gamma} \mathbb{A}_\gamma$.

For every pair of ordinals $\delta \leq \gamma$ there is a V-functor $\mathbb{A}_\gamma \xrightarrow{L^\gamma_\delta} \mathbb{A}_\delta$ such that $L^\gamma_\delta S_\gamma = S_\delta$ and such that for any triple of ordinals $\delta \leq \gamma \leq \beta$, $L^\gamma_\delta L^\beta_\gamma = L^\beta_\delta$ $(L^\gamma_\gamma = id)$

For every ordinal γ, $\mathbb{A}_{\gamma+1} = \mathbb{A}^{\mathbb{T}_\gamma}$ and $L^{\gamma+1}_\gamma = U^{\mathbb{T}_\gamma}$ where $\mathbb{T}_\gamma = (T_\gamma, \mu_\gamma, \eta_\gamma)$ is the codensity V-monad of S_γ, and $S_{\gamma+1} = \overline{S}_\gamma$ is the semantical comparison V-functor of S_γ .

For every limit ordinal β, $\mathbb{A}_\beta$ is the limit of the $\mathbb{A}_\gamma$'s for all $\gamma < \beta$ (in a sense that will be made precise in the subsequent proof of this proposition), L^β_γ are the projections of the limit and S_β is the V-functor corresponding to the S_γ's by the universal property of the limit.

b) For every ordinal γ, $\mathbb{A}_\gamma$ is V-complete and for every pair $\delta \leq \gamma$, L^γ_δ creates V-limits and cotensors (hence it is small V-continuous) and is V-faithful.

<u>Proof</u>:

Observe first that all three properties stated in b) are preserved under composition of V-functors. We do the proof by induction. $\mathcal{A}_0 = \mathcal{A}$ and $S_0 = R$. The non-limit step in the construction of the tower is clear. Statement b) follows from Propositions II.4.7, II.4.8 and the above observation.

Let β be a limit ordinal. $\mathcal{A}_\beta$ is the V-category whose objects are β-tuples $A_\beta = (A_\gamma)_{\gamma<\beta}$ $A_\gamma \in \mathcal{A}_\gamma$ such that for every pair of ordinals $\delta \leq \gamma < \beta$ $L_\delta^\gamma A_\gamma = A_\delta$, with a V-structure defined by:

$$\begin{array}{ccc} \mathcal{A}_\beta(A_\beta, B_\beta) & = & \varprojlim_{\gamma<\beta} \mathcal{A}_\gamma(A_\gamma, B_\gamma) \\ & {\scriptstyle L_\gamma^\beta} \searrow \quad \swarrow {\scriptstyle P_\gamma} & \\ & \mathcal{A}_\gamma(A_\gamma, B_\gamma) & \end{array}$$

where the limit is taken over the diagram

$$\mathcal{A}_\gamma(A_\gamma, B_\gamma) \xrightarrow{L_\delta^\gamma} \mathcal{A}_\delta(A_\delta, B_\delta) \; : \text{for every pair } \delta \leq \gamma < \beta \, .$$

(Observe that since the base functor preserves limits, $\mathcal{A}_\beta$ as a (ordinary) category is the limit of the $\mathcal{A}_\gamma$'s as (ordinary) categories.)

The required arrows:

$$\begin{array}{ccc}
\mathbb{A}_\beta(A_\beta, B_\beta) \otimes \mathbb{A}_\beta(B_\beta, C_\beta) & \dashrightarrow & \mathbb{A}_\beta(A_\beta, C_\beta) \\
\Big\downarrow {\scriptstyle L^\beta_\gamma \otimes L^\beta_\gamma} & & \Big\downarrow {\scriptstyle L^\beta_\gamma} \\
\mathbb{A}_\gamma(A_\gamma, A_\gamma) \otimes \mathbb{A}_\gamma(B_\gamma, C_\gamma) & \longrightarrow & \mathbb{A}_\gamma(A_\gamma, C_\gamma)
\end{array}$$

and

$$\begin{array}{ccc}
I & \dashrightarrow & A_\beta(A_\beta, A_\beta) \\
 & \searrow & \Big\downarrow {\scriptstyle L^\beta_\gamma} \\
 & & A_\gamma(A_\gamma A_\gamma)
\end{array}$$

are provided by the universality of the limit.

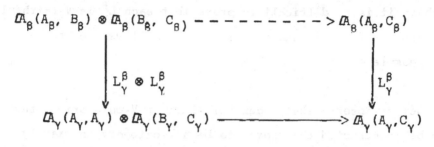

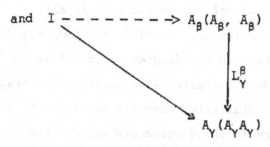

It is not difficult to check that the $\mathbb{A}_\beta$ defined above is actually a V-category and it is clear that the projections L^β_γ give a structure of V-functor to the functions $L^\beta_\gamma\ A_\beta = A_\gamma$ in such a way that the equation $L^\gamma_\delta\ L^\beta_\gamma = L^\beta_\delta$ holds. To check the universal property of $\mathbb{A}_\beta \xrightarrow{L^\beta_\gamma} \mathbb{A}_\gamma$, $\gamma < \beta$ is also straightforward. It follows then that there is a V-functor $\mathbb{C} \xrightarrow{S_\beta} \mathbb{A}_\beta$, unique such that $L^\beta_\gamma\ S_\beta = S_\gamma$. Finally, assuming statement b) for all pairs $\delta \leq \gamma < \beta$, from the definition (construction) of $\mathcal{A}_\beta$ and L_γ^β and the fact that ordinals form a chain it is not difficult to prove that each L_γ^β is V-faithful and creates V-limits and cotensors. It follows then that $\mathcal{A}_\beta$ is V-complete. ∎

Let us observe that a general limit of V-categories (that can be constructed and proven to be a V-category in exactly the same way as the above $\mathcal{A}_\beta$) is not V-complete even if all the V-categories in the diagrams are so. It will be V-complete if in addition all the V-functors in the diagram are small V-continuous. The projections do not create V-limits and cotensors but collectively create them. Similarly, they are not V-faithful even if all the V-functors in the diagram are so, but they are collectively V-faithful in the sense that the (canonical) inclusion into the product is V-faithful.

●● Proposition III.3.3

Given any ordinal α and any object $A_\alpha \in \mathcal{A}_\alpha$ such that $A_\alpha \xrightarrow{\eta_\alpha A_\alpha} T_\alpha A_\alpha$ is an isomorphism, then:

a) For every ordinal γ there exists a unique $A_\gamma \in \mathcal{A}_\gamma$ which is the given A_α when $\gamma = \alpha$, while for any pair $\delta \leq \gamma$, $L_\delta^\gamma A_\gamma = A_\delta$.

b) For any $\gamma \geq \alpha$ $A_\gamma \xrightarrow{\eta_\gamma A_\gamma} T_\gamma A_\gamma$ is an isomorphism and for any pair $\gamma \geq \delta \geq \alpha$ $\mathcal{A}_\gamma(A_\gamma,-) \xRightarrow{L_\delta^\gamma} \mathcal{A}_\delta(A_\delta, L_\delta^\gamma(-))$ is an isomorphism.

Proof:

It is clear that for $\gamma < \alpha$ we just define $A_\gamma = L^\alpha_\gamma A_\alpha$. For $\gamma \geq \alpha$ we proceed by induction.

Suppose $A_\gamma \xrightarrow{\eta_\gamma A_\gamma} T_\gamma A_\gamma$ is an isomorphism, then by Proposition II.4.5 we know there is a unique $A_{\gamma+1} \in \mathcal{A}_{\gamma+1}$ such that $L^{\gamma+1}_\gamma A_{\gamma+1} = A_\gamma$ and

$\mathcal{A}_{\gamma+1}(A_{\gamma+1}, -) \xrightarrow{L^{\gamma+1}_\gamma} \mathcal{A}_\gamma(A_\gamma, L^{\gamma+1}_\gamma(-))$ is an isomorphism. It follows from Proposition II.4.1 and Proposition I.4.2 that $L^{\gamma+1}_\gamma \eta_{\gamma+1} A_{\gamma+1}$ is an isomorphism, therefore, since $L^{\gamma+1}_\gamma$ reflects isomorphisms, $\eta_{\gamma+1} A_{\gamma+1}$ is an isomorphism. Finally, since $L^{\gamma+1}_\delta = L^\gamma_\delta L^{\gamma+1}_\gamma$, clearly $L^{\gamma+1}_\delta A_{\gamma+1} = A_\delta$ for any $\delta \leq \gamma+1$ and $\mathcal{A}_{\gamma+1}(A_{\gamma+1}, -) \xrightarrow{L^{\gamma+1}_\delta} \mathcal{A}_\delta(A_{\gamma+1}, L^{\gamma+1}_\delta(-))$ is an isomorphism for any $\gamma+1 \geq \delta \geq \alpha$.

Now let $\beta > \alpha$ be a limit ordinal, then $A_\beta = (A_\gamma)_{\gamma<\beta}$ is an object of $\mathcal{A}_\beta$, obviously unique, such that $L^\beta_\gamma A_\beta = A_\gamma$ for every $\gamma \leq \beta$. Let δ be any ordinal $\alpha \leq \delta < \beta$. For every $B_\beta \in \mathcal{A}_\beta$, $\mathcal{A}_\beta(A_\beta , B_\beta) \xrightarrow{L^\beta_\gamma} \mathcal{A}_\gamma(A_\gamma, B_\gamma)$, $\gamma < \beta$ is by definition a limit diagram, therefore, since

$\mathcal{A}_\gamma (A_\gamma, B_\gamma) \xrightarrow{L^\gamma_\delta} \mathcal{A}_\delta (A_\delta, B_\delta)$ is an isomorphism for every $\delta \leq \gamma < \beta$, it follows that $\mathcal{A}_\beta (A_\beta, B_\beta) \xrightarrow{L^\beta_\delta} \mathcal{A}_\delta(A_\delta B_\delta)$ is an isomorphism. So $\mathcal{A}_\beta(A_\beta, -) \xrightarrow{L^\beta_\delta} \mathcal{A}_\delta(A_\delta, L^\beta_\delta(-))$ is an

isomorphism. It follows then by exactly the same argument that in the jump of length one (Propositions II.4.1 and I.4.2) that $L^{\beta}_{\delta}\ \eta_{\beta}\ A_{\beta}$ is an isomorphism. In particular $L^{\beta}_{\alpha}\eta_{\beta}A_{\beta}$ is an isomorph since
If $\delta<\alpha$, $L^{\beta}_{\delta} = L^{\alpha}_{\delta}\ L^{\beta}_{\alpha}$, it is obvious that $L^{\beta}_{\delta}\ \eta_{\beta}\ A_{\beta}$ is also an isomorph

It is clear that the V-functors L^{β}_{δ}, $\delta < \beta$, collectively reflect isomorphisms, hence $\eta_{\beta}\ A_{\beta}$ is an isomorphism.

•• Proposition III.3.4 ($\mathbb{V}$ well-powered)

Given any collection (A_{γ}), $A_{\gamma} \in \mathcal{A}_{\gamma}$, ($\gamma$ ranging over the class (ordered category) of all ordinals) such that for every pair $\delta \leq \gamma$, $L^{\gamma}_{\delta}\ A_{\gamma} = A_{\delta}$, there is a α such that $A_{\alpha} \xrightarrow{\eta_{\alpha}A_{\alpha}} T_{\alpha}A_{\alpha}$ is an isomorphism.

Proof:

For every object $C \in \mathbb{C}$, $\mathcal{A}_{\gamma}(A_{\gamma},\ S_{\gamma}C)$ is a decreasing chain of sub-objects of $\mathcal{A}_{0}(A_{0},\ S_{0}C)$, and therefore ($\mathbb{V}$ is well-powered) it becomes stationary. Since $\mathbb{C}$ is small, there is an ordinal α such that $\mathcal{A}_{\alpha+1}(A_{\alpha+1},\ S_{\alpha+1}(-)) \xrightarrow{L^{\alpha+1}_{\alpha}} \mathcal{A}_{\alpha}(A_{\alpha}, S_{\alpha}(-))$ is an isomorphism. Then, by Proposition II.4.6 we are done.

In view of statement a) of Proposition III.3.3 and this last proposition, collections (A_{γ}) $A_{\gamma} \in \mathcal{A}_{\gamma}$ such that for every pair $\delta \leq \gamma$ $L^{\gamma}_{\delta}\ A_{\gamma} = A_{\delta}$, are completely determined by any single one of the objects A_{γ} for γ large enough. There is a canonical one, namely, the first one. It is possible then to treat such collections (proper classes) as sets, in particular, by abuse of language, we can form the class of all such collections. (Strictly speaking, this class would be the class whose elements are objects $\dot{A}_\alpha \in \mathbb{A}_\alpha$ for some α, such that $\eta_\alpha A_\alpha$ is an isomorphism and such that for every $\gamma < \alpha$, $\eta_\gamma A_\gamma$ ($A_\gamma = L_\gamma^\alpha A_\alpha$) is not an isomorphism). This class has a structure of V-category, it sits on top of all the $\mathbb{A}_\gamma$'s and it is a limit over all the ordinals of the $\mathbb{A}_\gamma$'s. Explicitly, the statement of Proposition III.3.2 can be prolonged one step more.

•• Proposition III.3.5 ($\mathbb{V}$ well-powered)

There is a V-complete V-category $\mathbb{B}$ and for every ordinal γ a small V-continuous V-faithful V-functor which creates V-limits and cotensors, $\mathbb{B} \xrightarrow{L_\gamma} \mathbb{A}_\gamma$, such that for any two ordinals $\delta \leq \gamma$, $L_\delta = L_\delta^\gamma L_\gamma$. $\mathbb{B}$ with projections L_γ is the limit over all the ordinals of the V-categories $\mathbb{A}_\gamma$.

It follows that there is a V-functor $\mathbb{C} \xrightarrow{S} \mathbb{B}$ such that $S_\gamma = L_\gamma S$ for every γ.

Proof:

$\mathbb{B}$ is the V-category whose objects are collections $A = (A_\gamma)$, $A_\gamma \in \mathbb{A}_\gamma$ such that $L_\delta^\gamma A_\gamma = A_\delta$, with a V-structure defined by:

$$\begin{array}{ccc} \mathbb{B}(AB) = \varprojlim_{\gamma} \mathbb{A}_\gamma(A_\gamma, B_\gamma) & & \\ \quad \searrow^{L_\gamma} & \swarrow_{p_\gamma} & \\ & \mathbb{A}_\gamma(A_\gamma, B_\gamma) & \end{array}$$

where the limit is taken over the (large) diagram:

$\mathbb{A}_\gamma(A_\gamma, B_\gamma) \xrightarrow{L^\gamma_\delta} \mathbb{A}_\delta(A_\delta, B_\delta)$, for every pair of ordinals $\delta \leq \gamma$.

The above limit exists (in $\mathbb{V}$) since the diagram (because every L^γ_δ is V-faithful) is a decreasing chain of sub-objects of $\mathbb{A}_0(A_0, B_0)$ and $\mathbb{V}$ is well powered. The rest of the statement follows exactly in the same way that the limit case in Proposition III.3.2. ∎

Observe that, if we had not been so careful about the set theoretical legitimacy of the class of objects of $\mathbb{B}$, Proposition III.3.5 would have followed directly without any need of Propositions III.3.3 and III.3.4. However, the non-trivial result is not the existence of $\mathbb{B}$ but the fact that the process under which the tower is built stops after $\mathbb{B}$. Namely, the codensity V-monad of $\mathbb{C} \xrightarrow{S} \mathbb{B}$ is the identity. Proving this result is what we have essentially done in Propositions III.3.3 and III.3.4.

●● Proposition III.3.6

The V-functor $\mathbb{C} \xrightarrow{S} \mathbb{B}$ is V-codense, that is, for every $A \in \mathbb{B}$, $A \xrightarrow{\eta A} TA$ is an isomorphism (where $\mathbb{T} = (T, \mu, \eta)$ is the codensity V-monad of S).

Proof:

From Proposition III.3.4 it follows that there is an ordinal α such that $\eta_\alpha A_\alpha$ is an isomorphism (where $A_\alpha = L_\alpha A$). The proof is completed then in exactly the same way that the limit step in Proposition III.3.3. ∎

●● Remark III.3.2

For any $A \in B$, there is an α such that $\eta_\alpha\ A_\alpha$ is an isomorphism, and for such an α $\mathbb{B}(A,-) \xrightarrow{L\alpha} \mathbb{A}_\alpha(A_\alpha, L_\alpha(-))$ is an isomorphism. ∎

The following are some informal considerations concerning the V-category $\mathbb{B}$ and the process under which it has been obtained. Starting with the codensity V-monad of the V-functor $\mathbb{C} \xrightarrow{R} \mathbb{A}$, we obtain its V-category of algebras, where we have the codensity V-monad of the lifting (semantical comparison V-functor), which in turn gives rise to a V-category of algebras. After going up in this way an infinite number of times, we have the limit of the diagram (chain) of V-categories so obtained. Since there is a V-functor from $\mathbb{C}$ into this limit, we have its codensity V-monad, and in this way we go up through all the ordinals. The V-category $\mathbb{B}$ is the limit of the tower so obtained. We can think that its objects are those objects of $\mathbb{A}$ which can be lifted all the way up, that is, which admit an structure of algebra at every level. More correctly, in view of the possibility of different liftings, the objects of $\mathbb{B}$ are objects of $\mathbb{A}$ together with a structure of algebra at every level. Since the lifted R, $\mathbb{C} \xrightarrow{S} \mathbb{B}$, is V-codense and the V-faithful V-functor $\mathbb{B} \xrightarrow{L_0} \mathbb{A}$ is V-continuous, any object of $\mathbb{A}$ that can be lifted all the way up (that is, which is the underlying object of an object of $\mathbb{B}$) is a small end (hence small V-limit, Proposition I.3.5) of cotensors of objects of the form RC $\in \mathbb{A}$, C $\in \mathbb{C}$. (Observe that an iterated cotensor is again a cotensor, namely, $\bar{\mathbb{A}}(V, \bar{\mathbb{A}}(W, RC)) \approx \bar{\mathbb{A}}(V \otimes W, RC)$). An adequate converse of this fact is also true. To simplify the matter, consider the ordinary set based case; then, any limit in $\mathbb{A}$ of a diagram in $\mathbb{C}$ is an object of $\mathbb{A}$ that can be lifted all the way up. (Since R has a lifting into every level, just take, for every α, the limit in $\mathbb{A}_\alpha$). We see then that $\mathbb{B}$ "consists" of all limits (over a diagram in $\mathbb{C}$) of objects of the form RC, C $\in \mathbb{C}$, with a structure of category other than that of full sub-category of $\mathbb{A}$. Namely, we take as morphisms of $\mathbb{B}$ only those maps of $\mathbb{A}$ which are morphisms of algebras all the way up.

●● Theorem III.3.2 (2º Relative completion)

Let $\mathbb{C}$ be any small V-category and $\mathbb{C} \xrightarrow{R} \mathbb{A}$ a V-full-and-faithful V-dense (hence V-continuous) V-functor into a V-complete V-category. Then $\mathbb{C} \xrightarrow{S} \mathbb{B}$ (as in Proposition III.3.6) is a V-full-and-faithful V-continuous and V-cocontinuous

V-codense and V-generating V-functor into a V-complete and V-cocomplete V-category.

<u>Proof</u>:

There is the factorization $\mathbb{C} \xrightarrow{R} \mathbb{A}$, $\mathbb{C} \xrightarrow{S} \mathbb{B} \xrightarrow{L_0} \mathbb{A}$.

Observe first that from the previous Proposition and Theorem III.2.3 it follows that L_0 has a V-left adjoint $\mathbb{A} \xrightarrow{F_0} \mathbb{B}$ given by the formula $F_0 = \mathrm{Ran}_R(S)$, and therefore, since R is V-full-and-faithful $F_0\, R \approx S$. Also $R \xRightarrow{\eta R} \mathrm{Ran}_R(R)R$ is an isomorphism.

S is V-full-and-faithful:

For any pair of objects $C, D \in \mathbb{C}$, consider the factorization $\mathbb{C}(C,D) \xrightarrow{R} \mathbb{A}(RC, RD)$, $\mathbb{C}(C,D) \xrightarrow{S} \mathbb{B}(SC, SD) \xrightarrow{L_0} \mathbb{A}(RC, RD)$.

Since $RC \xrightarrow{\eta RC} \mathrm{Ran}_R(R)(RC)$ is an isomorphism, it follows from Remark III.3.2 that L_0 is an isomorphism. So R being an isomorphism by assumption, the result follows.

$\mathbb{B}$ is V-complete and S is V-continuous:

We know already that $\mathbb{B}$ is V-complete. From the assumption that R is V-continuous and the fact that L_0 creates ends, V-limits and cotensors it follows that S is V-continuous.

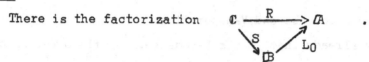

$\mathbb{B}$ is V-cocomplete and S is V-cocontinuous:

Since S is V-codense, the result follows from Propositions III.2.4 and III.1.8.

S is V-codense and V-generating:

We know already that S is V-codense. On the other hand, by assumption we have $\mathrm{Lan}_R(R) \approx \mathrm{id}$, hence $F_0L_0 \approx F_0\mathrm{Lan}_R(R)L_0 \approx \mathrm{Lan}_R(F_0R)L_0 \approx \mathrm{Lan}_R(S)L_0$. (Recall that since F_0 has a V-right adjoint it preserves left Kan extensions). Since L_0 is V-faithful, by Proposition 0.3 $F_0L_0 \Longrightarrow \mathrm{id}$ is a pointwise V-epimorphism, hence every object in $\mathbb{B}$ is a V-quotient-object of a coend of tensors of objects if the form $SC \in \mathbb{B}$ $C \in \mathbb{C}$. The result follows then from Remark II.3.1 dual. ∎

We finish this section with some observations aimed to give an additional insight into the tower constructed in Proposition III.3.2.

From Proposition II.4.4 it is not difficult to prove that every one of the V-functors $\mathbb{A}_\alpha \xrightarrow{L^\alpha_\beta} \mathbb{A}_\beta$ $\beta < \alpha$ has a V-left adjoint $\mathbb{A}_\beta \xrightarrow{F^\alpha_\beta} \mathbb{A}_\alpha$ given by the formula $F^\alpha_\beta = \mathrm{Ran}_{S_\beta}(S_\alpha)$. Furthermore, for any fixed β; all the V-monads in $\mathbb{A}_\beta$ determined by the pairs $F^\alpha_\beta \dashv_V L^\alpha_\beta$ are the same and are equal to $\mathbb{T}_\beta$, the codensity V-monad of S_β. ($L^\alpha_\beta F^\alpha_\beta = L^\alpha_\beta \mathrm{Ran}_{S_\beta}(S_\alpha) = \mathrm{Ran}_{S_\beta}(L^\alpha_\beta S_\alpha) = \mathrm{Ran}_{S_\beta}(S_\beta)$). Also, any "free" algegra is necessarily trivial in the next (hence in all the higher) level. That is, if $A_\alpha \in \mathbb{A}_\alpha$ is such that $A_\alpha = F^\alpha_\beta A_\beta$ for some $A_\beta \in \mathbb{A}_\beta$ $\beta < \alpha$, then $A_\alpha \xrightarrow{\eta_\alpha A_\alpha} T_\alpha A_\alpha$ is an isomorphism. It is clear also that every V-functor $\mathbb{B} \xrightarrow{L_\alpha} \mathbb{A}_\alpha$ has a V-left-adjoint $\mathbb{A}_\alpha \xrightarrow{F_\alpha} \mathbb{B}$, $F_\alpha = \mathrm{Ran}_{S_\alpha}(S)$, and for every object $A \in \mathbb{B}$ there is an ordinal α and an object $A_\alpha \in \mathbb{A}_\alpha$ such that $A = F_\alpha A_\alpha$. (Namely, $A_\alpha = L_\alpha A$ for any α such that $\eta_\alpha A_\alpha$ is an isomorphism; for the existence of such an α see Proposition III.3.6) The V-category $\mathbb{B}$ is then the union of the images of all the F_α's. From this, and the observation that any "free" algebra is trivial, it follows that $\mathbb{B}$ is the colimit of the diagram: $\mathbb{A}_\beta \xrightarrow{F^\alpha_\beta} \mathbb{A}_\alpha$, for all ordinals $\beta \leq \alpha$. If the V-functor $\mathbb{C} \xrightarrow{R} \mathbb{A}$ is V-full-and-faithful, then all the S_α's are V-full-and-faithful and the equation $F^\alpha_\beta S_\beta = S_\alpha$ holds (since S_β is V-full-and-faithful and $F^\alpha_\beta = \mathrm{Ran}_{S_\beta}(S_\alpha)$). Furthermore, for every pair $\beta < \alpha$, $\mathrm{Lan}_{S_\beta}(S_\alpha) \approx F^\alpha_\beta \mathrm{Lan}_{S_\beta}(S_\beta) L^\alpha_\beta$.

($\mathrm{Lan}_{S_\alpha}(S_\alpha) \xRightarrow{\approx} \mathrm{Lan}_{S_\beta}(S_\alpha)L^\alpha_\beta = \mathrm{Lan}_{S_\beta}(F^\alpha_\beta S_\beta)L^\alpha_\beta \approx F^\alpha_\beta \mathrm{Lan}_{S_\beta}(S_\beta)L^\alpha_\beta$. The first isomorphisms easily seen from the formulas provided by Theorem I.4.3 dual). In particular $\mathrm{Lan}_{S_\alpha}(S_\alpha) \approx F^\alpha_0 \mathrm{Lan}_R(R) L^\alpha_0$, hence if R is V-dense, $\mathrm{Lan}_{S_\alpha}(S_\alpha) \approx F^\alpha_0 L^\alpha_0$. It can be seen

that the whole density V-comonad of S_α (whose V-endofunctor is, by definition, $\mathrm{Lan}_{S_\alpha}(S_\alpha)$) is the one determined by the pair $F_0^\alpha \dashv_V L_0^\alpha$. Hence, since L_0^α is V-faithful, when R is V-full-and-faithful and V-dense, all the S_α's are V-generating. Finally, let us observe that similarly to the fact that $\mathbb{A} \xrightarrow{F_0} \mathbb{B}$ is $\mathrm{Ran}_R(S)$, $\mathbb{B} \xrightarrow{L_0} \mathbb{A}$ is $\mathrm{Ran}_S(R)$.
$(L_0 \approx L_0\, \mathrm{Ran}_S(S) \approx \mathrm{Ran}_S\,(L_0 S) = \mathrm{Ran}_S(R))$.

CHAPTER IV
FUNCTOR CATEGORIES

<u>Section 1</u> <u>V-functor categories</u>

In ordinary set-based category theory it is possible to form functor categories (with small exponent), small limits of categories and a variety of other constructions that, unlike the material previously developed in this paper, have conclusions of an existential character without any existential pre-assumptions in the starting data. Given any category $\mathbb{A}$ and a small category $\mathbb{C}$: <u>there is</u> a category $\mathbb{A}^{\mathbb{C}}$ of functors and natural transformations between them. In all cases as in this one, this is possible because of the basic existential axiom of set theory. As it is by now clear (after the work of Lawvere), this axiom translates into categorical language in completeness properties of the category of sets, that in turn, allow the development of set based category theory without recourse to the basic set-theoretical existential axiom. For example, given two functors $\mathbb{C} \underset{H}{\overset{T}{\rightrightarrows}} \mathbb{S}$, the set of natural transformations $\mathbb{S}^{\mathbb{C}}(T, H)$ can be produced as a (small) limit in the category of sets: Consider the comma category $(1, T)$ and the functor $(1, T) \xrightarrow{\Diamond} \mathbb{C}$ given by the rule $\Diamond(x \in TC) = C$. Then

$$\mathbb{S}^{\mathbb{C}}(T H) = \underset{\longleftarrow}{\lim} \left((1, T) \xrightarrow{\Diamond} \mathbb{C} \xrightarrow{H} \mathbb{S}\right) .$$

In other words; $\mathbb{S}^{\mathbb{C}}(T,H) = \mathrm{Ran}_T(H)(1)$.

Even in our non-autonomous treatment of V-vased category theory (that is, a V-category is a <u>class</u> of objects plus...., a small V-category is a <u>set</u> of objects plus.....) it is clear now that in order to have enough V-objects to give a V-structure to certain categorical constructions it will be necessary to assume completeness properties in $\mathbb{V}$ that will replace the basic set-theoretical existential axiom. (As was already the case with the need of equalizers for the V-structure of the category of algebras over a V-monad). Through this chapter then, $\mathbb{V}$ is a complete category, or equivalently, is a V-complete V-category.

Let $\mathbb{C}$, $\mathbb{A}$ be any V-categories, $\mathbb{C}$ small. The V-functor V-category $\mathbb{A}^{\mathbb{C}}$ is the category of V-functors and V-natural transformations with a V-structure given by: $\mathbb{A}^{\mathbb{C}}(T, H) = \int_C \mathbb{A}(TC, HC)$.

We have seen already in pages 57, 58 that $(\mathbb{A}^{\mathbb{C}})_o = \mathbb{V}_o(I, \mathbb{A}^{\mathbb{C}}(T,H))$ is the set of V-natural transformations between T and H. With the aid of the universal property of ends it is easy to check that this definition actually produces a V-category.

The projections $\mathbb{A}^{\mathbb{C}}(T, H) \xrightarrow{e_C} \mathbb{A}(TC, HC)$ give a V-functor structure to the evaluations functors: $\mathbb{A}^{\mathbb{C}} \xrightarrow{e_C} \mathbb{A}$, $e_C(T) = TC$.

There is a V-functor $\mathbb{A}^{\mathbb{C}} \otimes \mathbb{C} \xrightarrow{e} \mathbb{A}$ $e(T, C) = TC$, with a V-structure:

$$\mathbb{A}^{\mathbb{C}}(T, H) \otimes \mathbb{C}(C, D) \xrightarrow{e_C \otimes H}$$

$$\longrightarrow \mathbb{A}(TC, HC) \otimes \mathbb{A}(HC, HD) \xrightarrow{\circ} \mathbb{A}(TC, HD)$$

and for any other V-category $\mathbb{D}$ and any category $\boldsymbol{\Gamma}$; there is a one to one and onto correspondence between the arrows labeled with the same letter within each of the following two diagrams:

$$\begin{array}{ccc} \mathbb{D} \otimes \mathbb{C} & \xrightarrow{T} & \mathbb{A} \\ {\scriptstyle T\otimes id}\downarrow & \nearrow{\scriptstyle e} & \\ \mathbb{A}^{\mathbb{C}} \otimes \mathbb{C} & & \end{array}$$ commutative,

$$\begin{array}{ccc} \boldsymbol{\Gamma} \times \mathbb{C} & \xrightarrow{\Gamma} & \mathbb{A} \\ {\scriptstyle \Gamma \times id}\downarrow & & \nearrow{\scriptstyle e} \\ \mathbb{A}^{\mathbb{C}} \times \mathbb{C} & \longrightarrow & \mathbb{A}^{\mathbb{C}} \otimes \mathbb{C} \end{array}$$ commutative.

Here $\boldsymbol{\Gamma} \times \mathbb{C} \xrightarrow{\Gamma} \mathbb{A}$ is a functor such that for every $\lambda \in \boldsymbol{\Gamma}$, $\mathbb{C} \xrightarrow{\Gamma(\lambda,-)} \mathbb{A}$ is a V-functor and for every $\lambda \xrightarrow{f} \mu \in \boldsymbol{\Gamma}$, $\Gamma(\lambda,-) \Longrightarrow \Gamma(\mu,-)$ is a V-natural transformation.

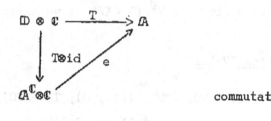

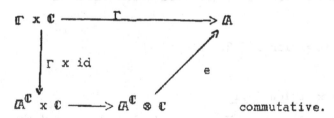

When $\mathbb{A} = \mathbb{V}$; the left and right Yoneda functors

$\mathbb{C} \xrightarrow{L} (\mathbb{V}^{\mathbb{C}})^{op}$ and $\mathbb{C} \xrightarrow{R} \mathbb{V}^{\mathbb{C}^{op}}$ $\quad LC = \mathbb{C}(C,-)$, $RC = \mathbb{C}(-,C)$

(that is, the left Yoneda is the right Yoneda of the dual category dualized) have a structure of V-functors

given by:

$$\begin{array}{ccc} \mathbb{C}(D,D') & \overset{R}{\dashrightarrow} & \mathbb{A}^{\mathbb{C}^{op}}(RD,\ RD') \\ & {\scriptstyle \mathbb{C}(C,-)} \searrow \quad \swarrow {\scriptstyle e_C} & \\ & \mathbb{V}(\mathbb{C}(C,D),\ \mathbb{C}(C,D')) & \end{array}$$

and for any V-functor $T \in \mathbb{V}^{\mathbb{C}}$ $(T \in \mathbb{V}^{\mathbb{C}^{op}})$ we have:

●● <u>Proposition IV.1.1</u>

$\mathbb{V}^{\mathbb{C}}(\mathbb{C}(C,-),\ T) = TC$. $\quad (\mathbb{V}^{\mathbb{C}^{op}}(\mathbb{C}(-,\ C),\ T) = TC)$, in particular, the Yoneda V-functors are V-full-and-faithful.

<u>Proof</u>:

Proposition I.5.2. ∎

●● <u>Proposition IV.1.2</u>

The left Yoneda V-functor is V-codense and the right Yoneda V-functor is V-dense.

<u>Proof</u>:

Formula (1) (page 59) means exactly that the dual of the left Yoneda is V-dense, that is, the left Yoneda is V-codense.

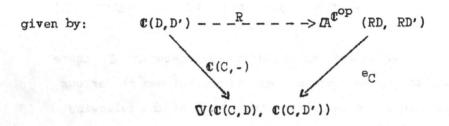

The same formula applied to the category $\mathbb{C}^{op}$ means that the right Yoneda is V-dense. ∎

As in the ordinary set base case, the following is true:

•• Proposition IV.1.3

For any V-category $\mathbb{A}$ and small V-category $\mathbb{C}$, if $\mathbb{A}$ is V-well-powered, then so is $\mathbb{A}^{\mathbb{C}}$. ∎

Also, as it would be expected, if $\mathbb{A}$ is a V-complete V-category so is $\mathbb{A}^{\mathbb{C}}$. More precisely:

•• Proposition IV.1.4

Given any V-category $\mathbb{A}$ and a small V-category $\mathbb{C}$, the family of evaluation V-functors, $\mathbb{A}^{\mathbb{C}} \xrightarrow{e_C} \mathbb{A}$, collectively creates V-limits, ends, cotensors and the respective dual concepts. In particular, if $\mathbb{A}$ is V-complete (V-cocomplete), so is $\mathbb{A}^{\mathbb{C}}$, and the evaluation V-functors are small V-continuous, (V-cocontinuous).

Proof:

All we have to do is to check that, when $\mathbb{C}$ is small, pointwise V-limits of V-functors, pointwise ends of V-functors and pointwise cotensors of V-functors are preserved by the representables $\mathbb{A}^{\mathbb{C}} \xrightarrow{\mathbb{A}^{\mathbb{C}}(T,\,-)} \mathbb{V}$, $T \in \mathbb{A}^{\mathbb{C}}$.

Let $\mathbb{\Gamma}$ any category, given $\dfrac{\mathbb{\Gamma} \times \mathbb{C} \xrightarrow{\Gamma} \mathbb{A}}{\mathbb{\Gamma} \xrightarrow{\Gamma} \mathbb{A}^{\mathbb{C}}}$,

(see page 151), suppose the pointwise V-limit of Γ exists. Then:

$$\mathbb{A}^{\mathbb{C}}\,(T,\ \underset{\overleftarrow{\lambda}}{\text{V-lim}}\,\Gamma(\lambda,-)) = \int_C \mathbb{A}(TC,\ \underset{\overleftarrow{\lambda}}{\text{V-lim}}\,\Gamma(\lambda,C)) =$$

$$= \int_C \underset{\overleftarrow{\lambda}}{\text{V-lim}}\ \mathbb{A}(TC,\Gamma(\lambda,C)) = \text{(formula (1) page 38)}$$

$$= \underset{\overleftarrow{\lambda}}{\text{V-lim}}\ \int_C \mathbb{A}(TC,\ \Gamma(\lambda,\ C)) = \underset{\overleftarrow{\lambda}}{\text{V-lim}}\,\mathbb{A}^{\mathbb{C}}(T,\ \Gamma(\lambda,\ -))$$

Exactly in the same way, (but using formula (1) page 36), it can be checked for ends. For cotensors:

$$\mathbb{A}^{\mathbb{C}}\,(T,\ \overline{\mathbb{A}^{\mathbb{C}}}(V,H)) = \int_C \mathbb{A}(TC,\ \overline{\mathbb{A}}(V,\ HC)) \approx \int_C \mathbb{V}(V,\ \mathbb{A}(TC,\ HC)) =$$

$$= \mathbb{V}(V,\ \int_C \mathbb{A}(TC,\ HC)) = \mathbb{V}(V,\ \mathbb{A}^{\mathbb{C}}(T,\ H))\ .$$

Finally, it is clear that the pointwise structures are completely characterized by the fact that the canonical map: e_C (structure) $\longrightarrow$(structure)is the equality for every $C \in \mathbb{C}$.

∎

IV.1.4

Of the same nature as Proposition / is the fact that, given any V-complete V-category $\mathbb{A}$, and a V-functor $\mathbb{C} \xrightarrow{S} \mathbb{D}$, $\mathbb{C}$, $\mathbb{D}$ small, the process of taking right Kan extensions along $\mathbb{S}$ is actually a V-functor $\mathbb{A}^{\mathbb{C}} \xrightarrow{Ran_S} \mathbb{A}^{\mathbb{D}}$, V-right adjoint to the process (now a V-functor) of composing with S on the right, $\mathbb{A}^{\mathbb{D}} \xrightarrow{\mathbb{A}^S} \mathbb{A}^{\mathbb{C}}$. Effectively (see Day Kelly [1]); if $T \in \mathbb{A}^{\mathbb{D}}$, $H \in \mathbb{A}^{\mathbb{C}}$,

$$\mathbb{A}^{\mathbb{D}}(T,\ Ran_S(H)) = \int_D \mathbb{A}(TD, \int_C \bar{\mathbb{A}}(\mathbb{D}(D, SC), HC)) =$$

$$= \int_D \int_C \mathbb{A}(TD, \bar{\mathbb{A}}(\mathbb{D}(D, SC), HC)) =$$

$$= \int_C \int_D \mathbb{V}(\mathbb{D}(D, SC), \mathbb{A}(TD, HC)) =$$

$$= \int_C \mathbb{V}^{\mathbb{D}^{op}}(\mathbb{D}(-, SC), \mathbb{A}(T(-), HC)) =$$

$$= \int_C \mathbb{A}(TSC, HC) = \mathbb{A}^{\mathbb{C}}(TS, H) = \mathbb{A}^{\mathbb{C}}(\mathbb{A}^S(T), H).$$

By the way, notice that when $\mathbb{A} = \mathbb{V}$, the following is true: $\mathbb{V}^{\mathbb{C}}(T, H) = Ran_T(H)(I)$. In this case, the above V-adjointness is just the equality:

$$Ran_T(Ran_S(H))(I) = Ran_{TS}(H)(I) .$$

Section 2 The V-completion of a small V-category

Given any small V-category $\mathbb{C}$, we exhibit here two V-completions of $\mathbb{C}$ in the sense explained at the beginning of Section 3, Chapter III. Assume $\mathbb{V}$ is complete and well-powered (hence V-complete and V-well-powered).

●● Theorem IV.2.1 (first V-completion)

Let $\mathbb{B}$ the V-full sub-category of $\mathbb{V}^{\mathbb{C}^{op}}$ whose objects are all V-continuous V-subfunctors of small ends (hence small V-limits) of cotensors of representables. Then $\mathbb{B}$ is V-complete and V-cocomplete, the right Yoneda V-functor $\mathbb{C} \xrightarrow{R} \mathbb{V}^{\mathbb{C}^{op}}$ factors through $\mathbb{B}$, $\mathbb{C} \xrightarrow{S} \mathbb{B}$, and S is V-full-and-faithful, V-continuous and V-cocontinuous, V-dense and V-cogenerating.

Proof:

First observe that from the V-Yoneda Lemma it follows that V-continuous V-functors $\mathbb{C}^{op} \xrightarrow{T} \mathbb{V}$ are exactly those V-functors T for which $\mathbb{C}^{op} \xrightarrow{\mathbb{V}^{\mathbb{C}^{op}}(R(-),\ T)} \mathbb{V}$ is V-continuous.

The result follows then from Remark III.3.1 and Theorem III.3.1.b) once we observe that by Propositions IV.1.1,2,3,4, the right Yoneda V-functor $\mathbb{C} \longrightarrow \mathbb{V}^{\mathbb{C}^{op}}$ satisfies the hypotheses in Theorem III.3.1.b). ∎

●● Theorem IV.2.2 (Second V-completion)

Let $\mathbb{C} \xrightarrow{S} \mathbb{B}$ be as in Theorem IV.3.2 with respect to the right Yoneda V-functor $\mathbb{C} \xrightarrow{R} \mathbb{V}^{\mathbb{C}^{op}}$, then $\mathbb{B}$ is a V-codense V-generating completion. ∎

Recalling the considerations made in page 143, the V-completion offered in the above theorem consists of V-functors $\mathbb{C}^{op} \longrightarrow \mathbb{V}$ which are ends (hence V-limits) of co-tensors of representables with a structure of V-category other than that of V-full sub-category of $\mathbb{V}^{\mathbb{C}^{op}}$.

In the ordinary set-based case, this completion consists of exactly all limits of representables (since the right Yoneda is full-and-faithful they are necessarily limits of a diagram in $\mathbb{C}$) with less than all V-natural transformations between them. (Observe that because of the possibility of different liftings, we can have many non-isomorphic copies of the same functor $\mathbb{C}^{op} \longrightarrow \mathbb{V}$; they are determined by different diagrams which have the same limit in $\mathbb{V}^{\mathbb{C}^{op}}$).

At the end of the next section, we will observe that the second completion is contained (as a V-full-sub-category) in the dual of the first completion. (Observe that both are V-codense and V-generating).

<u>Section 3. The cosingular and corealization V-functors</u>

Given any V-functor from a small V-category $\mathbb{C}$ into a V-complete V-category $\mathbb{A}$, $\mathbb{C} \xrightarrow{S} \mathbb{A}$, S is always tractable and there is a V-full-and-faithful V-functor from the clone of operations of S into $(\mathbb{V}^{\mathbb{C}})^{op}$ $\mathbb{K}_S \xrightarrow{i} (\mathbb{V}^{\mathbb{C}})^{op}$,
$i(A) = \mathbb{A}(A, S(-))$ with a V-structure given by the identity.

The cosingular functor of S, $\mathbb{A} \xrightarrow{F} (\mathbb{V}^{\mathbb{C}})^{op}$ is the composite $\mathbb{A} \xrightarrow{F^S} \mathbb{K}_S \xrightarrow{i} (\mathbb{V}^{\mathbb{C}})^{op}$, $F(A) = \mathbb{A}(A, S(-))$ with a V-structure given by:

$$\begin{array}{ccc} \mathbb{A}(A, B) & \overset{F}{\dashrightarrow} & \mathbb{V}^{\mathbb{C}}(FB, FA) \\ & {\scriptstyle \mathbb{A}(-, SC)} \searrow \quad \swarrow {\scriptstyle e_C} & \\ & \mathbb{V}(\mathbb{A}(B, SC), \mathbb{A}(A, SC)) & \end{array}$$

The fact that $\mathbb{A}$ is V-complete implies that the V-right adjoint U^S of F^S can be extended into a V-functor $(\mathbb{V}^{\mathbb{C}})^{op} \xrightarrow{U} \mathbb{A}$, the corealization V-functor of S. It is defined on objects:

$$U(T) = \int_C \bar{\mathbb{A}}(TC, SC) \text{ for } \mathbb{C} \xrightarrow{T} \mathbb{V}.$$

and the following chain of V-natural isomorphisms implies that U is a V-functor V-right adjoint to F.

$$\mathbb{V}^{\mathbb{C}}(T, FA) = \int_C \mathbb{V}(TC, \mathbb{A}(A, SC)) \approx \int_C \mathbb{A}(A, \bar{\mathbb{A}}(TC, SC)) \approx$$
$$\approx \mathbb{A}(A, \int_C \bar{\mathbb{A}}(TC, SC)) = \mathbb{A}(A, UT).$$

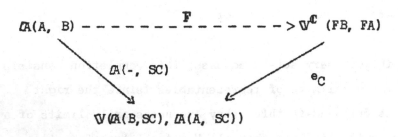

From the definition we have $UFA = Ran_S(S)(A) = \int_C \bar{\mathbb{A}}(\mathbb{A}(A, SC), SC$ and it can actually be checked that the whole V-monad determined by the pair $F \dashv_V U$ is the codensity V-monad of S, that is, $\mathbb{T}_U = \mathbb{T}_S$.

•• Proposition IV.3.1

In the diagram:

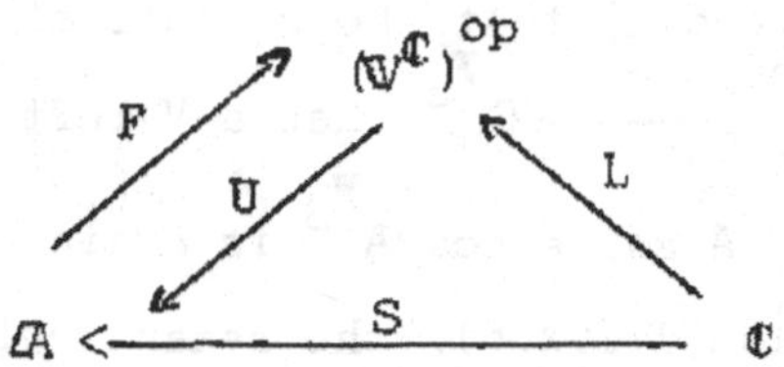

$F = Ran_S(L)$ and $U = Ran_L(S)$. $UL = S$ and if S is V-full-and-faithful, $FS = L$.

Proof:

$Ran_L(S)(T) = \int_C \overline{\mathbb{A}}(\mathbb{V}^{\mathbb{C}}(\mathbb{C}(C, -), T), SC) =$ (V-Yoneda lemma) $=$

$= \int_C \overline{\mathbb{A}}(TC, SC) = U(T).$

$Ran_S(L)(A) = \int_C \overline{(\mathbb{V}^{\mathbb{C}})^{op}}(\mathbb{A}(ASC), \mathbb{C}(C, -)) = \int^C \mathbb{A}(A, SC) \otimes_{\mathbb{V}^{\mathbb{C}}} \mathbb{C}(C,-) =$

$= \int^C \mathbb{A}(A, SC) \otimes \mathbb{C}(C, -) \approx \int^C \mathbb{C}(C,-) \otimes \mathbb{A}(A, SC) =$ (by Proposition I.5.1) $= \mathbb{A}(A, S(-)) = FA.$

Observe that the above chain of isomorphisms, read from bottom to top, prove that $Ran_S(L)$ exists. (Since we are not assuming $\mathbb{V}$ to be cocomplete, $(\mathbb{V}^{\mathbb{C}})^{op}$ is not V-complete and hence we cannot assume that $\mathrm{Ran}_S(L)$ exists).

Finally, by Proposition I.4.5, the rest of the statement is clear. ∎

Since the V-category $(\mathbb{V}^{\mathbb{C}})^{op}$ is V-cocomplete it follows from Theorem I.A (Appendix) that the semantical comparison V-functor of U, $(\mathbb{V}^{\mathbb{C}})^{op} \xrightarrow{\bar{U}} \mathbb{A}^{\mathbb{T}_S}$ has a V-left adjoint $\mathbb{A}^{\mathbb{T}_S} \xrightarrow{\bar{F}} (\mathbb{V}^{\mathbb{C}})^{op}$. Also, since $\mathbb{A}^{\mathbb{T}_S}$ is again a V-complete V-category (Proposition III.2.6), the semantical comparison V-functor of S, $\mathbb{C} \xrightarrow{\bar{S}} \mathbb{A}^{\mathbb{T}_S}$ gives rise to another pair of V-adjoint functors between the same categories $(\mathbb{A}^{\mathbb{T}_S} \rightleftarrows (\mathbb{V}^{\mathbb{C}})^{op})$: its cosingular and corealization V-functors. These two pairs of V-adjoint functors are actually the same. Effectively, let $\bar{G}$ be the corealization V-functor of $\bar{S}$, we have the diagram:

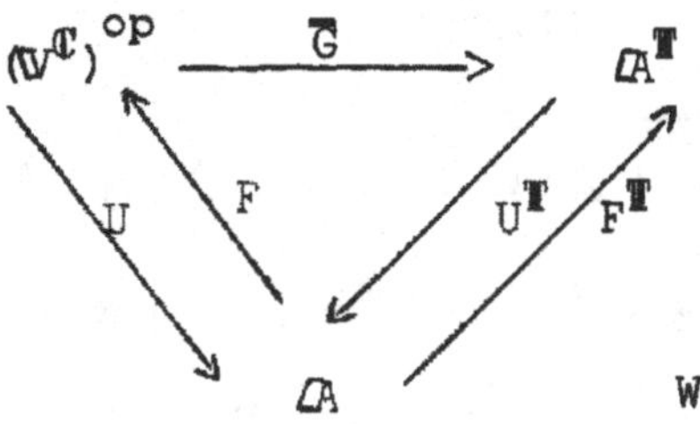

Where $\mathbb{T} = \mathbb{T}_U = \mathbb{T}_S$

$\bar{G}FA = \int_C \overline{\mathbb{A}^{\mathbb{T}}}(\mathbb{A}(A, SC), \bar{S}C) = \mathrm{Ran}_{\bar{S}}(S)(A)$ which in turn, by Proposition II.4.4 is equal to $F^{\mathbb{T}}$. On the other hand,

$U^{\mathbb{T}}\bar{G}(T) = U^{\mathbb{T}} \int_C \overline{\mathbb{A}^{\mathbb{T}}}(TC, \bar{S}C) = \int_C \mathbb{A}(TC, SC) = U(T)$. So it follows from Proposition II.1.6 that $\bar{G} = \bar{U}$, which implies that the whole adjoint pair is the same.

We gather together all the information that we have about the above situation in our next proposition.

•• Proposition IV.3.2

Given a V-functor $\mathbb{C} \xrightarrow{S} \mathbb{A}$ from a small V-category into a V-complete V-category, it gives rise to the following diagram of V-categories and V-functors:

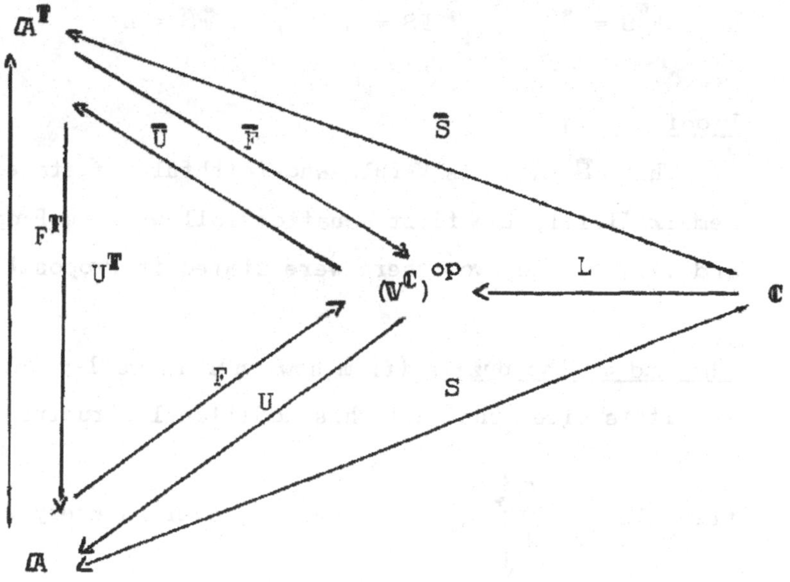

where $\bar{S}$ is the semantical comparison V-functor of S, F, $\bar{F}$, U, $\bar{U}$ are the cosingular and corealization V-functors of S and $\bar{S}$ respectively. $\bar{U}$ also is (both are equal) the semantical comparison V-functor of U. The following equations hold:

$$UL = S \quad , \quad \bar{U}L = \bar{S} \quad , \quad U^{\mathbb{T}}\bar{S} = S$$

$$\bar{U}F = F^{\mathbb{T}} \quad \text{and} \quad U^{\mathbb{T}}\bar{U} = U \quad .$$

■

•• Remark IV.3.1

With the same situation as in the above proposition, if S is V-full-and-faithful then $\bar{S}$ also is V-full-and-faithful and the following additional equations hold:

$$F^{\mathbb{T}}S = \bar{S} \quad , \quad FS = L \quad , \quad \bar{F}\bar{S} = L$$

Proof:

That $\bar{S}$ also is V-full-and-faithful was stated in Remark II.1.1, the first equation follows from Propositions I.4.5 and II.4.4, the two others were stated in Proposition IV.1.1.

■

The end of the puzzle (from now on $\mathbb{V}$ is well-powered)

It is clear that all this additional structure "inside" the triangle

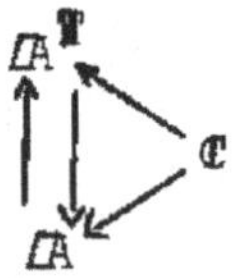

carries over in every step in the construction of the tower (Proposition III.3.2). That is, given a V-functor $\mathbb{C} \xrightarrow{R} \mathbb{A}$ from a small V-category into a V-complete V-category, let, for every ordinal γ, $\mathbb{C} \xrightarrow{S_\gamma} \mathbb{A}_\gamma$ be as in Proposition III.3.2. Then we have:

(1)

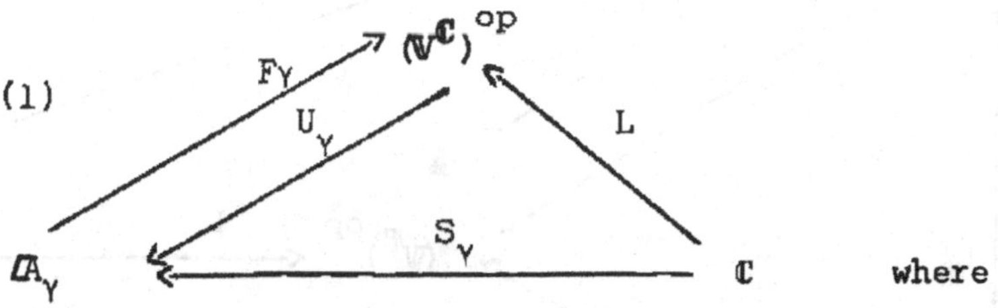

where F_γ and U_γ are the cosingular and corealization V-functors of S_γ, $F_\gamma \dashv_V U_\gamma$, and $F_\gamma = \mathrm{Ran}_{S_\gamma}(L)$, $U_\gamma = \mathrm{Ran}_L(S_\gamma)$, $U_\gamma L = S_\gamma$ and if R (hence S_γ) is V-full-and-faithful, $F_\gamma S_\gamma = L$.

For a limit ordinal β, U_β is also the V-functor $(\mathbb{V}^{\mathbb{C}})^{op} \longrightarrow \mathbb{A}_\beta$ determined by the V-functors $(\mathbb{V}^{\mathbb{C}})^{op} \xrightarrow{U_\gamma} \mathbb{A}_\gamma$ $\gamma < \beta$. In effect, write U'_β this V-functor. Since each of the U_γ's is V-continuous, it follows that U'_β is also V-continuous. Also; since for every $\gamma < \beta$; $U_\gamma L = S_\gamma$, it follows that $U'_\beta L = S_\beta$. Hence $U'_\beta L = U_\beta L$, and so, since both are V continuous and L is V-codense (Proposition IV.1.2) it easily follows that they are equal.

For every pair of ordinals $\delta < \gamma$, we have: (see considerations at the end of Section 3, Chapter III).

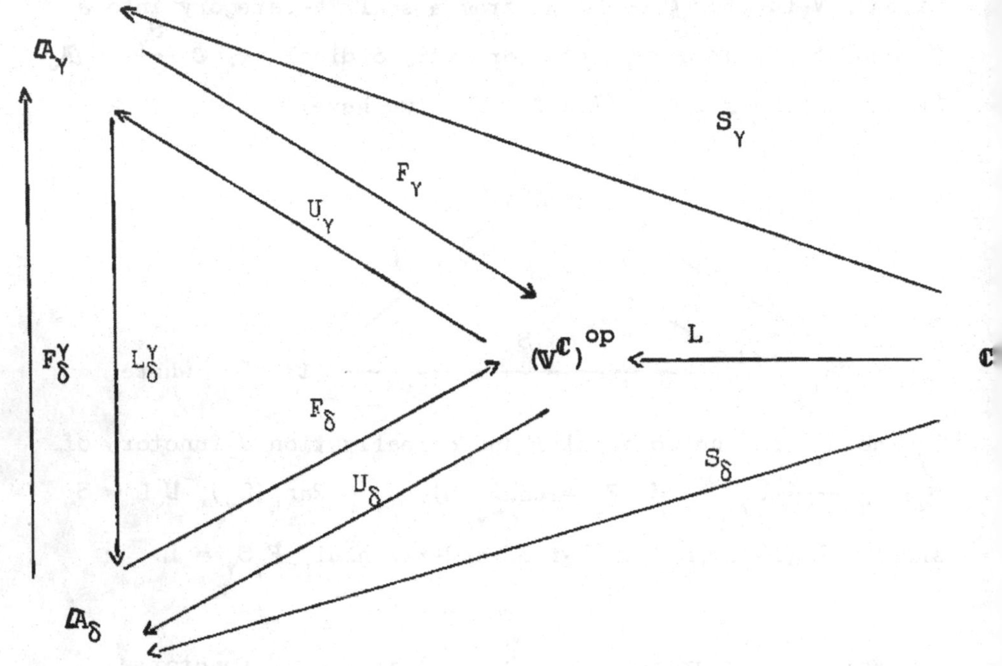

$L^\gamma_\delta\, S_\gamma = S_\delta$, $F^\gamma_\delta \dashv_V L^\gamma_\delta$, $F^\gamma_\delta = \mathrm{Ran}_{S_\delta}(S_\gamma)$, $U_\gamma\, F_\delta = F^\gamma_\delta$ and $L^\gamma_\delta\, U_\gamma = U_\delta$ (see also equations below diagram (1) page 163).

If $\mathbb{R}$ (hence S_δ) is V-full and faithful, $F^\gamma_\delta\, S_\delta = S_\gamma$.

Considering both ends of the tower we have:

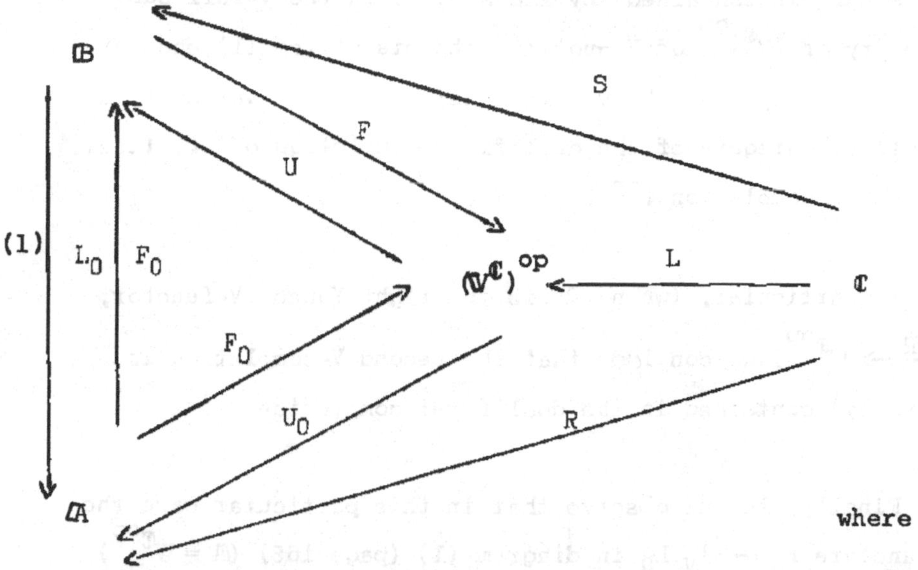

where

in addition we have that F (the cosingular V-functor of S) is V-full-and-faithful and $L_0 = \mathrm{Ran}_S(R)$.

When R is V-dense and V-full-and-faithful, S is V-generating, and from the equation FS = L, since the "V-inclusion" F is V-cocontinuous, it follows that every object B of 𝔹 (considered as a V-functor ℂ ⟶ 𝕍, that is, as an object of $(\mathbb{V}^{\mathbb{C}})^{op}$) is a V-quotient object (in $(\mathbb{V}^{\mathbb{C}})^{op}$) of a coend of tensors of representables. Namely, since $\mathrm{Lan}_S(S)(B) \longrightarrow B$ is a V-epimorphism, $\mathrm{Lan}_S(L)(B) = \mathrm{Lan}_S(FS)(B) = F\,\mathrm{Lan}_S(S)(B) \longrightarrow FB$ is a V-epimorphism. Since S is V-continuous, B considered as a V-functor ℂ ⟶ 𝕍 is also V-continuous, (FB = 𝔹(B, S(-)), and so $\mathbb{B}$ is contained (by means of F) in the V-full subcategory of $(\mathbb{V}^{\mathbb{C}})^{op}$ of V-quotient objects of (small) ends of tensors of representables. That is; $\mathbb{B}$ is V-equivalent to a V-full subcategory of the dual first completion of $\mathbb{C}$, (i.e., the first completion $\mathbb{C}^{op}$).

In particular, (when R is the right Yoneda V-functor, $\mathbb{C} \xrightarrow{R} \mathbb{V}^{\mathbb{C}^{op}}$) we conclude that the second V-completion is (V-fully) contained in the dual first completion.

Finally, let us observe that in this particular case the V-functors $F_0 \dashv_V L_0$ in diagram (1) (page 165) ($\mathbb{A} = \mathbb{V}^{\mathbb{C}^{op}}$) are the singular and realization V-functors of S, (denoting by L_0' the singular V-functor of S, we have $L_0'S = R$ (dual of Proposition IV.3.1). Since we always have $L_0S = R$, the equation $L_0' = L_0$ follows from the V-codensity of S and the fact that both are V-continuous). ∎

APPENDIX

The notion of cotensor is independent of the notion of V-limit, but it is clear by now that cotensors behave in all respects as if they were V-limits. In introducing cotensors in this paper (actually, in introducing the dual concept, Chapter I Section 2) we made the (trivial) observation that in the ordinary set-based world this is actually the case. Explicitly, given any object A in a (locally small) category $\mathbb{A}$ and any set S, the formula (1) $\bar{\mathbb{A}}(S,A) = \prod_S A$ holds. This formula holds only in the set-based context, but for many closed categories $\mathbb{V}$, it is still the case that the concept of cotensor is not independent of that of the limit. That is, for those $\mathbb{V}$, cotensors are real limits, and can be constructed by means of a similar but more complicated formula generalizing formula (1) above.

In this appendix we give general conditions which (when satisfied by a closed category $\mathbb{V}$) imply that in the V-based world cotensors are real limits. More explicitly, if a closed category $\mathbb{V}$ satisfies these conditions, then any V-category $\mathbb{A}$ which has small limits is cotensored, and cotensors are constructed in terms of limits by means of a specific formula. Clearly, the same conditions imply the dual result, that is, any V-category $\mathbb{A}$ which has small colimits is tensored, and tensors are constructed in terms of colimits by means of a specific formula.

Basic in proving this result is a theorem of [8] that we now state in its generalized V-version.

Theorem A.1.

Given a V-adjoint triangle,

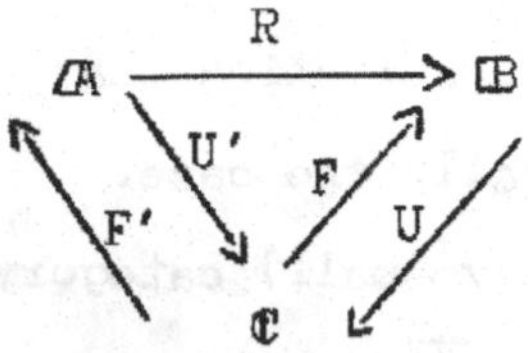

$$(\epsilon, \eta) : F \dashv_V G$$

$$(\epsilon', \eta') : F' \dashv_V G'$$

(the vertices V-categories and the arrows V-functors, UR = U') such that the diagram:

$$FUFU \underset{\epsilon FU}{\overset{FU\epsilon}{\rightrightarrows}} FU \xrightarrow{\epsilon} id$$

is a V-coequalizer of V-functors, then, if it exists, the following coequalizer (1) of V-functor is a V-left adjoint $\mathbb{B} \xrightarrow{L} \mathbb{A}$ of $\mathbb{A} \xrightarrow{R} \mathbb{B}$.

$$F'UFU \overset{F'U\epsilon}{\Longrightarrow} F'U \Longrightarrow L$$

(1) $$F'UFU \xrightarrow{F'U\theta U} F'URF'U = F'U'F'U \xrightarrow{\epsilon' F'U} F'U$$

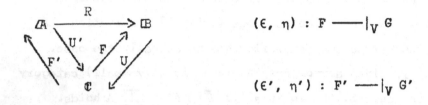

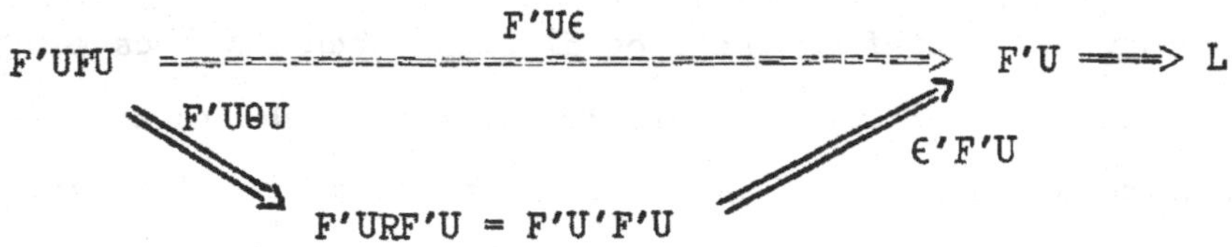

$$\theta = (F \xrightarrow{F\eta'} FU'F' = FURF' \xrightarrow{\epsilon RF'} RF') \quad .$$

The proof given in [8] translates word by word into this general V-context, and so we do not give a proof here. ∎

Notice that the V-coequalizer (1) will exist (pointwise) if $\mathbb{A}$ has V-coequalizers. More generally, it would be enough to assume that $\mathbb{A}$ has V-coequalizers of reflexive pairs. (the double arrow $F'U \xrightarrow{F'\eta U} F'UFU$ is a reflection for the pair of double arrows in (1)).

Now we record explicitly an observation needed to prove our next result.

Observation A.1

Let $\mathbb{V}$ be a closed category such that the base functor $\mathbb{V} \xrightarrow{\mathbb{V}_o(I,-)} \mathbb{S}$ reflects isomorphisms. Then, given any V-functor $\mathbb{B} \xrightarrow{G} \mathbb{A}$ ($\mathbb{B}$, $\mathbb{A}$ any V-categories), a functor $\mathbb{A} \xrightarrow{F} \mathbb{B}$, left adjoint to G, $(\theta_o, \epsilon, \eta) : F \longrightarrow\!| \; G$, has a structure of V-functor V-left adjoint to G.

Proof:

For any given $A \in \mathbb{A}$ (fixed) define:

$$\theta = (\mathbb{B}(FA, -) \xrightarrow{G} \mathbb{A}(GFA, G(-)) \xrightarrow{\mathbb{A}(\eta A, \square)} \mathbb{A}(A, G(-)))$$

It is clear that θ is V-natural. On the other hand, it is immediate that for any $B \in \mathbb{B}$, $\mathbb{V}_o(I, \theta B) = \theta_o AB$, so θ is an isomorphism. The result follows then from Proposition 0.2. ∎

If $\mathbb{V}$ is a closed category with small coproducts, then the base functor $\mathbb{V} \xrightarrow{\mathbb{V}_o(I, -)} \mathbb{S}$ has a left adjoint

$$S \xrightarrow{-\otimes_{\mathbb{V}} I} \mathbb{V},\ (S \otimes_{\mathbb{V}} I = \coprod_{S} I),\ \mathrm{id} \overset{\eta}{\Longrightarrow} \mathbb{V}_o(I, -\otimes_{\mathbb{V}} I),$$

$$\mathbb{V}_o(I, -)\otimes_{\mathbb{V}} I \overset{\epsilon}{\Longrightarrow} \mathrm{id} \ .$$

<u>Theorem A.2</u>

Let $\mathbb{V}$ be a closed category with small coproducts such that for every $V \in \mathbb{V}$ the diagram:

$$\mathbb{V}_o(I, \mathbb{V}_o(I,V)\otimes_{\mathbb{V}} I)\otimes_{\mathbb{V}} I \underset{\epsilon\mathbb{V}_o(I,V)\otimes_{\mathbb{V}} I}{\overset{\mathbb{V}_o(I,\ \epsilon V)\otimes_{\mathbb{V}} I}{\rightrightarrows}} \mathbb{V}_o(I,V)\otimes_{\mathbb{V}} I \xrightarrow{\epsilon V} V$$

is a coequalizer in $\mathbb{V}$, then

a) Any V-category $\mathbb{A}$ with small colimits is tensored, moreover, given any $A \in \mathbb{A}$ and $V \in \mathbb{V}$, there are explicitly determined maps such that the following diagram is a coequalizer in $\mathbb{A}$.

$$\text{(1.a)} \quad \mathbb{V}_o(I, \mathbb{V}_o(I,V) \otimes_{\mathbb{V}} I)\otimes_{\mathbb{A}} A \rightrightarrows \mathbb{V}_o(I,V)\otimes_{\mathbb{A}} A \longrightarrow V \otimes_{\mathbb{A}} A\ ,$$

Where $S \otimes_{\mathbb{A}} A = \coprod_{S} A$ (for any set S).

b) Dually, any V-category $\mathbb{A}$ with small limits is cotensored, moreover, given any $A \in \mathbb{A}$ and $V \in \mathbb{V}$, there are explicitly determined maps such that the following diagram is an equalizer in $\mathbb{A}$.

$$\bar{\mathbb{A}}(V,A) \longrightarrow \bar{\mathbb{A}}_o(\mathbb{V}_o(I,V),A) \rightrightarrows$$

$$\rightrightarrows \bar{\mathbb{A}}_o(\mathbb{V}_o(I, \mathbb{V}_o(I,V) \otimes_{\mathbb{V}} I), A)$$

where $\bar{\mathbb{A}}_o(S,A) = \prod_S A$ (for any set S).

<u>Proof</u>:

a) Consider the adjoint triangle:

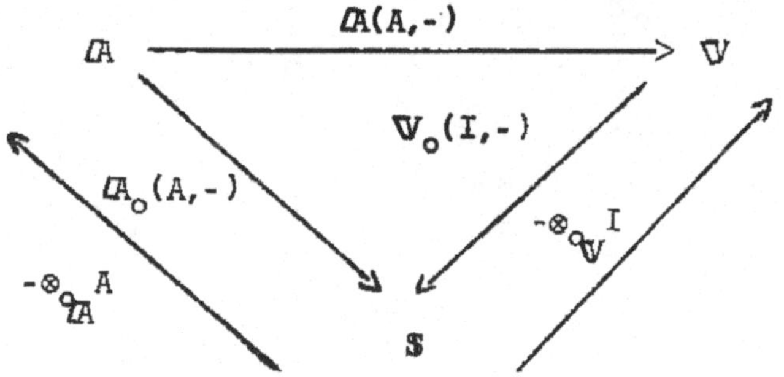

then, by the set-based version of Theorem A.1 $\mathbb{A}(A,-)$ has a left adjoint which is computed as the coequalizer (1.a) above. It only remains to see then that this left adjoint is actually a V-left adjoint. This follows from Observation A.1 once we notice that the conditions imposed in $\mathbb{V} \xrightarrow{\mathbb{V}_o(I,-)} \mathbb{S}$ imply that $\mathbb{V}_o(I, -)$ reflects isomorphisms.

Part b) is just part a) applied to the V-category $\mathbb{A}^{op}$ (which is cocomplete). ∎

Finally, let us remark that it is possible to see (using Theorem A.1 for example) that the condition imposed on $\mathbb{V}$ in Theorem A.2 implies that $\mathbb{V}$ is a full reflexive sub-category of the category of algebras over the monad in S determined by the pair of adjoint functors $- \otimes_{\mathbb{V}} I \dashv \mathbb{V}_o(I, -)$.

BIBLIOGRAPHY

[1] B. J. DAY and G. M. KELLY, Enriched functor categories, Lecture Notes Vol. 106, Springer Verlag.

[2] S. EILEMBERG and G. M. KELLY, Closed categories, Proc. Conf. on Categorical Algebra. (LaJolla 1965), Springer Verlag.

[3] G. M. KELLY, Adjunction for enriched categories, Lecture Notes Vol. 106, Springer Verlag.

[4] J. LAMBEK, Completions of Categories, Lecture Notes Vol. 24, Springer Verlag.

ADDITIONAL REFERENCES

[5] H. APPELGATE and M. TIERNEY, Categories with models, Lecture Notes Vol. 80, Springer Verlag.

[6] J. BENABOU, Criteres de representabilite des foncteurs, C. R. Acad. Sc. Paris, t. 260, p. 752-755 (1965).

[7] M. BUNGE, Relative Functor Categories and Categories of Algebras, Journal of Algebra Vol. II, Jan. 1969, p. 64-101.

[8] E. DUBUC, Adjoint Triangles, Lecture Notes Vol. 61, Springer Verlag.

[9] F. E. J. LINTON, An outline of functorial semantics, Lecture Notes Vol. 80, Springer Verlag.

[10] F. ULMER, On cotriple and Andre (co) Homology, their relationship with classical Homological algebra, Lecture Notes Vol. 80, Springer Verlag.

Offsetdruck: Julius Beltz, Weinheim/Bergstr.